Danielly Vieira de Lucena
Luciana Viana Amorim
Carlos Magno Souto

Inhibited fluids for shale drilling

Danielly Vieira de Lucena
Luciana Viana Amorim
Carlos Magno Souto

Inhibited fluids for shale drilling

An experimental study

ScienciaScripts

Imprint
Any brand names and product names mentioned in this book are subject to trademark, brand or patent protection and are trademarks or registered trademarks of their respective holders. The use of brand names, product names, common names, trade names, product descriptions etc. even without a particular marking in this work is in no way to be construed to mean that such names may be regarded as unrestricted in respect of trademark and brand protection legislation and could thus be used by anyone.

Cover image: www.ingimage.com

This book is a translation from the original published under ISBN 978-613-9-68467-0.

Publisher:
Sciencia Scripts
is a trademark of
Dodo Books Indian Ocean Ltd. and OmniScriptum S.R.L publishing group

120 High Road, East Finchley, London, N2 9ED, United Kingdom
Str. Armeneasca 28/1, office 1, Chisinau MD-2012, Republic of Moldova, Europe
Printed at: see last page
ISBN: 978-620-8-21822-5

SUMMARY

Acknowledgements 3
SUMMARY 5
Chapter 1 6
Chapter 2 10
Chapter 3 30
Chapter 4 42
Chapter 5 90
BIBLIOGRAPHICAL REFERENCES 92

"Without dreams, life has no sparkle. Without goals, dreams have no foundation. Without priorities, dreams don't become real. Dream, set goals, establish priorities and take risks to make your dreams come true. It's better to err by trying than to err by omitting." *(Augusto Cury)*

*To my parents **Damiao** and **Socorro**, and to my brother **Danyllo**, for their encouragement, love and unconditional support at all times, especially in times of uncertainty, which are very common for those trying to tread new paths in life.*

I dedicate this work

CURRICULUM VITAE

Materials Engineer at the Federal University of Campina Grande, UFCG.
Period: 2004-2008.

MSc in Materials Science and Engineering from the Federal University of Campina Grande, UFCG.

Period: 2009-2011.

Professor at the Federal Institute of Paraiba

Thanks

0imited space for acknowledgments brings with it the equally difficult challenge of writing this work, and so it certainly doesn't allow me to thank, as I should, all the people who, in the course of this work, have helped me to fulfill my objectives and achieve another desired stage in my academic training. I therefore leave you with just a few words, but with a sincere and deep sense of gratitude.

Above all, I thank God for the opportunity to make this dream come true and, above all, for the determination I have been given to do so.

To my parents Damiao and Socorro, for all your love and unconditional support in my decisions. I thank you for making it possible for me to do this work through your immeasurable support and sacrifice, for giving me encouragement in times of discouragement. You will have my eternal gratitude for the moments when you were by my side, making me believe that I am capable.

To my brother Danyllo, for a lifetime of living together and sharing, who was there not only to teach me, but above all with his understanding and endless friendship throughout all these years of work.

To my boyfriend, Carlos Magno, who in recent times has made a decisive contribution by encouraging and supporting me throughout the completion of this work, I thank you for your love, encouragement and companionship at this crucial time.

To Prof. Dr. Hélio de Lucena Lira, for his competence, academic stimulation, theoretical contributions, support, relevant suggestions and teachings.

To the participants of the examining board, Prof. Dr. César, Prof. Dr. Gelmires de Araùjo Neves, Prof. Dr. José Agnelo Soares and Researcher Renalle Cristina Alves de Medeiros Nascimento, for their contributions.

To the Postgraduate Coordinators, Prof. Dr. Gelmires de Araùjo Neves and Prof. Dr. Romualdo, for all their attention and support during the thesis.

I would also like to thank my new family, the Irmaos da Luz, in the form of Felipe, Gabriela, Gutemberg, Italo, Kamila, Kelma, Klebson, Lumena, Natalia, Raphaela, Romulo, Valkir, Vanusa, Wendel and William, who have given me a new outlook on life, I thank you immensely for all your strength and prayers. Without you, I wouldn't have been able to finish this thesis with my heart so full of joy.

To my friends in the fluids group at LABDES, I would like to thank you for your pleasant company and for your constant help in carrying out this work. In particular, to my advisor Luciana Viana Amorim, who proved to be much more than an advisor, for her contributions, advice, for always guiding me along this path with her always impeccable conduct, thank you very much for your dedication, corrections, patience, understanding and fundamental motivation for the preparation of this thesis.

To Prof. Gelmires de Araújo Neves for supplying the sample of reactive clay. To our partner company System Mud - A Imdex Limited Company Indùstria e Comércio Ltda., for supplying the additives and for their technical assistance, especially to Technical Director Eugênio Pereira and Chemist Juliano Magalhaes, for their promptness, suggestions and information that made this work possible.

I would like to thank the IFPB, especially the Cajazeiras Campus, for having made it possible for me to do this work, and for their recognition and contribution to this important professional achievement.

Finally, to all those who in one way or another helped or accompanied me on this journey, thank you very much!

SUMMARY

A large part of the problems faced during the drilling of oil wells is associated with the instability caused by the interaction between clay formations and the fluid used to pass through them. Studying geological formations that are susceptible to hydration is a challenge because it is a phenomenon that is responsible for around 90% of the problems related to drilling oil wells, and because little is known about the mechanisms that govern this phenomenon, as well as because of the scarcity of studies aimed at analyzing different products that prevent the occurrence of hydration problems (inhibitors). The aim of this work was to study and evaluate the effectiveness of inhibited and chlorine-free aqueous drilling fluids in controlling hydration in oilfield formations in various regions of the country. Thirteen samples of Brazilian shales and two samples of bentonite clays were studied. Initially, all the samples were characterized in order to identify those most susceptible to hydration. Next, a study was carried out to select the best concentration of inhibitor to contain the expansion of reactive clays and, based on the results, the optimum concentration of inhibitor was established (20g/350mL of water). From this, drilling fluids were developed with four different expansive clay inhibitors (potassium sulphate, potassium acetate, potassium citrate and potassium chloride) and the pH, density, rheological and filtration properties were determined. The dispersibility of the fluids developed was also determined. Based on the results, it was concluded that drilling fluids have been successfully developed with satisfactory rheological, filtration and hydration control performances. Excellent results have also been obtained in terms of dispersibility rates. In general, the results obtained indicate that the potassium citrate inhibitor showed the best control of the reactivity of reactive formations and that it is an alternative product for replacing inhibitors already known in the industry.

Keywords: Reactive formations, shales, expansion inhibitors.

Chapter 1

1 INTRODUCTION

The problems of well stability, related to the increase in the number of drilling scenarios involving reservoirs, in which the presence of geological formations that are difficult to drill, such as expansive shales, is common, is one of the main difficulties encountered in the operation of oil wells. The great compositional variation at short distances and along the depth of the well also influence this context of drilling difficulties.

The problems that generate instability have traditionally been well managed through the use of oil-based drilling fluids, as they create an ideal barrier that controls the influx of water or ions into the formation and have advantages over water-based fluids, especially with regard to excellent swelling control and well stability, lubricity and protection against corrosion (Shuixiang *et al.*, 2011). However, the increased demands of environmental agencies in recent years have made the use of this type of fluid increasingly restricted (Dye *et al.*, 2006; Montilva, 2007; Silva *et al.*, 2011).

Water-based drilling fluids are increasingly being used for oil and gas exploration, as they are generally considered to be more environmentally acceptable than oil-based or synthetic fluids (Anderson *et al.*, 2010). Unfortunately, despite polluting less than other types of fluids, their use facilitates the hydration and swelling of the clay, generating physico-chemical problems that are accentuated by the non-ideal membrane behavior of the shales when in the presence of these fluids (ISMAIL AND HUANG, 2009).

According to Stefan (1956), when reactive formations are suspended in water, they adsorb a considerable amount of liquid, with an apparent increase in volume, by increasing the basal interplanar distance of the clay layers. This phenomenon is known as swelling or expansion. The magnitude of this phenomenon depends on the nature of the exchangeable cations and the structure of the surrounding medium (Villar *et al.*, 2012). The consequences of expansion range from the collapse of the walls and widening of the well to its complete closure. Thus, the stability of oil wells has been studied considering the mechanical and chemical aspects of the rock, in relation to fluid-rock interactions (MUNIZ *et al.*, 2005; RICHARD *et al.*, 2010, LUCENA *et al.*, 2011a).

According to Lopes *et al.* (2012), bentonite clay often appears as a component of expansive formations and causes problems when drilling oil wells. When it comes into contact with the water-based fluid, it hydrates and can reduce the diameter of the well and have the clay fragments dispersed and incorporated into the fluid itself, contaminating it in such a way that important rheological parameters, as well as the specific weight and volume of the filtrate, are altered.

The movement of water inside the shales occurs as a result of the gradient of chemical activity between the drilling fluids and the shales, due to osmotic pressure. Chemical activity defines the salinity of a drilling fluid, so a drilling fluid with high activity has low salinity, causing water to flow into the formation. In this way, the presence of salts (expansion inhibitors) in the drilling fluid is essential for maintaining the stability of reactive formations (RABE & CHERREZ, 2009).

According to Blachier (2009), swelling inhibitors must not only significantly reduce the hydration of reactive formations, but also meet increasingly stringent environmental guidelines. According to the definition proposed by Qu *et al.* (2009), swelling inhibitors are chemical compounds capable of reducing the swelling of reactive formations caused by contact with aqueous fluids.

According to Magalhaes (2012), the oil well drilling market uses a rich variety of inhibitor options for composing inhibited fluids. These inhibitors act on the clay's ionic/polar active sites located in the basal spacing and on the lateral edges, making it difficult for water to enter and effectively reducing the hydration of the clay (SERRA E SANTOS, 2004; BASSI *et al.*, 2009).

Research has shown that the interactions between shale and drilling fluid can be regulated to increase and improve well stabilization, and the results indicate that the best performances were shown by inhibitors with K ions$^+$ (VAN OORT, 2003, ISMAIL AND HUANG, 2009 and ANDERSON *et al.*, 2010).

Several studies have concluded that the use of techniques such as X-ray diffraction, differential thermal analysis and cation exchange capacity are important tools for explaining the swelling control promoted by inhibitors (ZAMPORI *et al.*, 2012; KHODJA *et al.*, 2010; SUTER *et al.*, 2009; DIAZ- PEREZ *et al.*, 2007; WARR AND BERGER, 2007; KATTI AND KATTI, 2006; PETIT, 2006).

For a better understanding of the interaction phenomena between the rock and the drilling fluid, it is first necessary to have a complete description of these active formations, both from the point of view of their individual constituents and their microstructure (Rabe and Fontoura, 2003). Therefore, the complete characterization of reactive formations is indispensable for a better understanding of the hydration phenomenon that occurs in this type of formation.

As such, there is a great need for research into the development and improvement of inhibited drilling fluids in view of the problems caused by the swelling of reactive formations in the oil well drilling process. As a result, the research group at UFCG's PEFLAB (Drilling Fluids Research Laboratory) has been developing characterization, evaluation and application studies for drilling fluids that meet these new demands. This work therefore aims to research inhibited aqueous drilling fluids that are suitable for drilling and susceptible to the swelling phenomenon.

1.1 Objectives

1.1.1 General

The aim of this work is to study and evaluate the efficiency of inhibited and chlorine-free aqueous drilling fluids in controlling the hydration of shale formations from various regions of the country.

1.1.2 Specifics

To carry out this work, the following specific objectives are proposed:

• to characterize systemically and mineralogically the samples of shales and bentonite clays to be studied;

• evaluating the efficiency of chlorine-free chemical inhibitors in controlling the hydration and consequent swelling of hydratable clays;

• analyze the rheological and filtration behavior of drilling fluids developed with and without chemical inhibition;

• evaluate the inhibition capacity of drilling fluids by means of the dispersibility test in the presence of highly reactive shales and,

• Correlate the inhibition power of the fluids developed with the characterization obtained for each of the reactive shales analyzed, and Correlate the dispersibility containment for non-reactive shales with the characterization obtained for them.

1.2 Work Organization

This work is organized into five (5) chapters. Chapter 1 presents an introduction, describing the importance and motivating factors for carrying out this study and its objectives.

Chapter 2 contains a literature review, starting with a description of clays and shales. The problem of the hydration of clays by drilling fluids is also addressed in this chapter. Definitions, functions and characteristics of drilling fluids are presented, as well as fluid-rock interaction, instability mechanisms, and the action and importance of expansion inhibitors in active geological formations.

Chapter 3 describes the materials selected and the methodology used to carry out this work. The methodology includes the characterization of the samples studied, the selection of the best concentrations of swelling inhibitors, the preparation of the fluids, rheological and filtration tests. The dispersibility test, used to assess the swelling inhibition capacity of the fluids, is also discussed.

Chapter 4 presents and discusses the results obtained from the clay and shale characterization data (cation exchange capacity, specific area, X-ray fluorescence, differential thermal analysis, thermogravimetry and X-ray diffraction), as well as swelling graphs, capillary suction time and bentonite swelling, rheological and filtration tests, and dispersibility results.

Chapter 5 presents the conclusions, highlighting the most positive results in relation to the proposed objectives. This is followed by suggestions for future work and the bibliography used as a reference for this work.

Chapter 2

---

2 LITERATURE review

2.1 Moisturizing formations

2.1.1 Bentonite clays

Bentonite is a fine-grained clay made up of montmorillonite-type clay minerals from the smectite group, consisting of two tetrahedral sheets of silica (SiO_2), separated by an octahedral sheet of alumina (Al_2O_3), joined together by oxygen common to the sheets. In the octahedral positions, the cation Al^{3+} can be replaced by Mg^{2+} , Fe^{3+} , and in the tetrahedral positions there can be isomorphic substitutions of Si^{4+} by Al^{3+} (Souza Santos, 1992). The structure of montmorillonite is illustrated in Figure 1.

Figure 1 - Structure of two montmorillonite lamellae.

Source: Cunha, 2013.

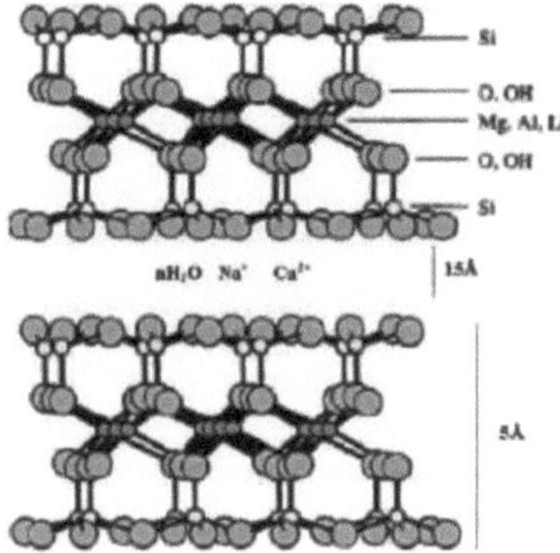

According to Menezes *et al.* (2009), smectitic clay minerals are characterized by their high degree of isomorphic substitution within their crystallographic structure, which leads to an excess of negative charge that is compensated by interlamellar cations.

Bentonite clays are included in the class of minerals of greatest industrial interest, with a wide range of applications (Amorim, 2005). They can be classified according to the predominant type of exchangeable cation. According to Ye *et al.* (2010), bentonites are characterized by their high cation exchange capacity and large specific area.

According to Suguio (2003), bentonite clays are also called expanded network clays because their layers are easily separated by adsorbed water. This characteristic is related to the ease with which clays belonging to the smectite group swell when placed in water.

From a geological point of view, bentonite is a mineral formed primarily by the weathering of volcanic rocks and largely composed of montmorillonite-type smectites. It has a great capacity to absorb water and swell. The origin of bentonites is related to the processes of weathering, erosion and deposition from volcanic rocks, lake and deep sea environments (GRIMM AND GÜVEN, 1978; VELDE, 1985; WEAVER, 1989; MEUNIER, 2005).

Smectite particles have a bipolar character, with negative charges located in the larger plane (face) and positive charges in the smaller region (edge), as shown in Figure 2.

Figure 2- Morphological representation of a colloidal smectite particle.

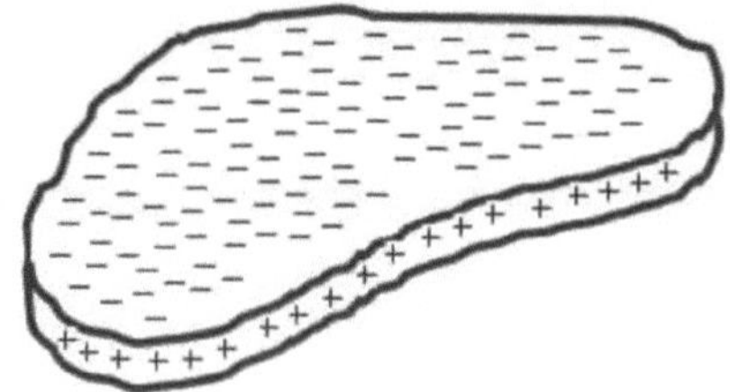

According to Cygan (2009), clay expansion in the presence of water is a very complex phenomenon. Studies of water adsorption by bentonite layers show that hydration capacity depends strongly on the exchangeable cations present in them, as well as on the water/clay ratio.

According to Santos (2001), the content of smectite present in shales is often the only property of the shale studied to define the type of drilling fluid chosen, due to the high expansibility potential of these clay minerals with their reactivity potential. Thus, knowledge of the properties of smectites is of great importance in explaining the phenomena related to the hydration of formations containing them in large proportions.

2.1.2 Leaflets

According to Fontoura (2012), among sedimentary rocks, shales are the most finely laminated and have different stratified layers with moderate to high clay content. He concluded that due to these characteristics, shales are susceptible to different phenomena that cause instability in the process of drilling oil and gas wells.

According to Suguio (2003), its fissile structure is partly formed by compaction and the existence of placoid (micaceous) minerals. In another publication, Suguio (1998) defines fissility as a property related to the ability of sediments to separate into planes.

According to Silva (2005), the composition of shales can vary according to the predisposition of the source rock, controlling the color, which varies from brownish red to black. According to Suguio (2003), shales usually derive from two types of environment: marine (rich in chlorite and clays from the illite group) or lacustrine (enriched with smectite).

According to Al-Bazali *et al.* (2009), shales play an important role in oil exploration and production and are commonly found as source rocks.

O' Brien and Chenevert (1973) associate the characteristics of shales, such as resistance, tendency to hydrate and dispersion, with the clay fraction and the type of clay mineral, i.e. shales that are more hydratable and more easily dispersed probably have considerable levels of reactive clay in their mineral composition. According to the authors, by characterizing the minerals that make them up, it is possible to know the potential instability of the shales in the face of drilling.

Manohar (1999) observed that the distinctive characteristics of shales are their clay content and their low pore connectivity. These characteristics make them susceptible to phenomena such as hydration and swelling, thus reducing the stability of the well.

According to Rabe and Fontoura (2003), in relation to the oil industry, shales are detrital sedimentary rocks formed by the densification and compaction of fine-grained sediments (clays and siltstones), where the clay fraction can vary from 15 to close to 100% (Rabe and Fontoura, 2003).

According to Van Oort (2003), the stability of clay-rich shales is profoundly affected by their complex physical and chemical interactions with drilling fluids. Thus, these rocks with a high clay content have shown significant alterations, such as expansion or swelling, when placed in contact with aqueous fluids due to the adsorption of polar water molecules or hydrated ions solubilized in the medium. These changes can cause the rock to collapse during drilling with water-based fluids (MACHADO, 2002, LUCENA *et al.*, 2011b).

Silva (2005) studied the lithology of the dark shales and found it to be rich in organic matter, in the process of fossilization associated with environments that alternated periods of drought and rain.

As these clay minerals have different hydration energies, the ability of a certain rock to adsorb water is a function of the types and quantities of clay minerals that make it up (MACHADO, 2002).

According to a study by Melédez (2010), the main problems associated with rock formations made up of shales are: well enlargement, well closure and tool entrapment, torque and obstruction, loss of circulation of casing drilling fluid and profiling.

According to Nascimento *et al.* (2010a), mechanical entrapment of the drill string is caused by obstruction or physical restriction and occurs when the drill string is in motion and fluid circulation is impeded. The factors that contribute to mechanical entrapment are: poor well cleaning, well geometry, accumulation of gravel in the annular space, collapse of well walls and sedimentation of large particles carried by the fluid, caused, for example, by the incorporation of clay into the fluid. Figure 3 illustrates the mechanical clamping of pipes.

Figure 3: Drilling column stuck due to clay swelling.

Source: Nascimento *et al.*, 2010a.

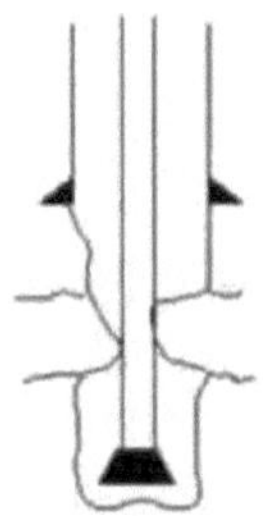

2.2 Reactive formations - drilling fluid interaction

During the drilling of oil and gas wells, various types of shale are encountered. According to Durand *et al.* (1995), shales are found in 75% of the formations drilled, and among the problems related to drilling, in 90% of cases shales are responsible for the instability of wells. These problems stem largely from the fact that the absorption of water in shale promotes the expansibility of the well and causes problems such as column entrapment caused by collapse. For Zhong *et al.* (2011), maintaining well stability is one of the most critical aspects of drilling.

According to Ismail and Huang (2009), the interaction between the aqueous drilling fluid and the water-sensitive zone, especially shale formations, is one of the main problems in the oil and gas industry, causing time and cost losses.

Reactive formations are prone to alterations in their structure in the presence of water, which can lead to major deformations. Billions of dollars are spent every year due to the damage caused by the excessive swelling of certain formations with a certain degree of reactivity (Keller, 2008; Pappala and Cerato, 2009; Powell *et al.*, 2013). In the United States alone, the damage is in the order of 500 million dollars a year. Reactive formations are those that show significant changes in volume due to the presence of moisture (ERZIN AND EROL,

2007).

The study carried out by Yang *et al.* (2011) indicates that this problem has become greater due to the increasing use of aqueous fluids, since oil-based fluids present restrictions due to the pollution of their residues and they conclude that despite polluting less than other fluids, commonly used in drilling reactive formations, the physico-chemical problems have become more accentuated due to the non-ideal membrane behavior of the shales when in the presence of these fluids.

According to Van Oort *et al.* (2003), the problem of well stability in expansive formations has frustrated oil field engineers since the beginning of oil and gas drilling operations. According to Osissanya *et al.* (2009), this problem may be the result of interactions between the shales and the drilling fluids and the existence of unfavorable stresses (formation pressure) in the rock or the result of a combination of these two processes.

According to Rabe and Fontoura (2003) and Amorim *et al.* (2007), the process of aqueous fluid-reactive formation interaction is the result of physical, chemical and mechanical phenomena that occur during and after drilling. This interaction can change the magnitude of the stresses in the formation around the well, hydrate the clay minerals and increase the moisture content of the formation, which can lead to the loss of tools and even the closure of the well.

Last and Plum (1995) attribute the problems related to the lack of stability to the swelling of expansible clays and shales and propose the use of chemical products that inhibit the hydration and expansibility of shales as a solution. Among these products are saline solutions, which try to reduce ionic migration through chemical equilibrium between rocks and fluids.

For some time, the stability of oil wells has been studied considering the mechanical and chemical aspects of the rock, the latter being related to fluid-rock interactions (Silva *et al.*, 2011). The swelling of shale particles when in contact with aqueous fluids has been considered the main cause of tool entrapment, due to the adsorption of water molecules or hydrated ions solubilized in the medium. In general, the content of hydratable clays in shale is over 50%. The stability of the well walls is a function of the rock-fluid interaction. As previously mentioned, there are chemical additives capable of minimizing these interactions. These additives lodge in the basal space, attaching themselves to the surface of the silicate sheets, preventing the entry of water molecules (MACHADO, 2002).

For Rabe and Cherrez (2009), understanding the transport phenomena and the behavior of the shale in relation to these fluids has proved to be one of the greatest challenges facing the oil industry due to the complexity of the phenomena surrounding these interactions.

2.2.1 Properties of clay minerals

2.2.1.1 Cation and anion exchange capacity

Clay minerals have the ability to exchange ions, i.e. they have ions fixed on the surface, between the layers and within the channels of the crystalline reticulum. This cation exchange capacity arises from the high potential of clays to react with cations present in solutions because they have negative charges on their external surface. These charges arise due to the isomorphic substitutions of Al^{3+} for Mg^{2+} and Si^{4+} for Al^{3+} , as they have broken bonds with the crystal's surface ions, as well as the hydrogen substitution of the hydroxyls (RABE, 2009).

According to Ye *et al.* (2010), the ionic radii of the ions are of great importance because the smaller the ion, the better they fit into the spaces in the water molecule clusters and these ions with small ionic radii have an attractive effect on the clay minerals to form an oriented structure.

Grim (1992) reports that the cation exchange capacity occurs differently in each type of clay mineral. In smectites, it is mainly due to isomorphic substitutions in the tetrahedral layer; in kaolinite, it is mainly due to broken bonds, and increases with the lower crystallinity of this clay mineral. In the case of illites and chlorites, this capacity is due to broken bonds and the exchange of K^+ on the edges of illites or Mg^{2+} on the surface of chlorites.

The exchangeable cations are held around the lateral edges of the clay mineral particles and, in the case of smectites, on the basal planes themselves. The exchangeable cations are electrostatically fixed along the faces and between the layers, due to the structural imbalance resulting from isomorphic substitutions and broken chemical bonds along the edges of the particles.

With regard to anion exchange capacity, this is still little known and has no defined experimental methodology. It is due to the instability of some clay minerals in the course of many anion exchange chemical reactions (GRIM, 1968).

According to Rabe (2003), anion exchange may be the result of broken bonds at the edges of the clay mineral particles. The author also indicates that the ability of some anions to react with clay minerals is related to a favorable geometric fit of the structure of these ions to the crystalline lattice units of the clay minerals.

Rabe *et al.* (2009), concluded that minerals with high values of cation exchange capacity (CEC) and specific surfaces have higher hydration capacities; in other words, higher capacities to adsorb water and therefore higher expansion rates.

According to Al-Bazali *et al.* (2006), CTC is a measure of the intensity of the negative charge surrounding the clay layers and is therefore the ability of the shales to act as semi-permeable membranes, based on the electrical exclusion of counter-ions (anions). The authors concluded after tests that the membrane efficiency of the shales increases when the CTC is also increased.

Zhang *et al.* (2008) presented a combination of the effects of CTC and permeability in measuring the membrane efficiency of shales. After subjecting different types of shale (Pierre, Arco-China, C1 and C2) to various solutions (NaCl, KCl, CaCl2 and KCOOH) at different water activities (0.93 and 0.85), the authors concluded that membrane efficiency is directly proportional to the CTC/water activity ratio and that high CTC/water activity values correlate well with high membrane efficiency values.

According to Ismail and Huang (2009), the reactivity of the shale is a function of the type and quantity of clay minerals present in the formation to be drilled. Based on this, they carried out a study using cation exchange capacity (CEC) to provide information for classifying the sensitivity of shales and concluded that the shale with the highest smectite content in its composition also had the highest CEC value.

2.2.1.2 Hydration of clays and shales

The drilling fluid comes into contact with the rock formation, which can be made up of minerals and rocks that are sensitive to water (Albuquerque, 2011). The factors associated with the instability of formations are related to the entry of water into the interlayers of clays (such as smectite), resulting in a "swelling" of the rock, as shown in Figure 4.

Figura 4- Structure and hydration of calcic and sodic montmorillonite.

Source: Lummus and Azar, 1986.

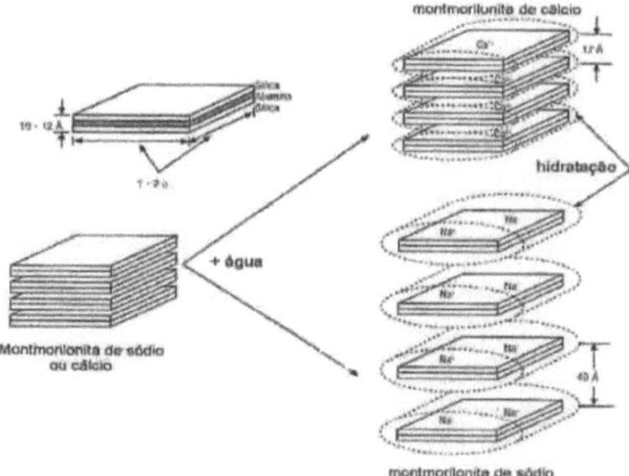

The clays are organized in laminar packets and have a high degree of hydration. When they come into contact with water, these clay sheets separate as the water penetrates the basal space (Figure 5). According to Pereira (2007), clay formations containing smectite are especially sensitive to the presence of water. Many of these formations contain various types and different amounts of clay. The greater the presence of smectite, the greater the reactivity in the presence of water.

Figura 5- Schematic representation of the layered delamination of a hydrated clay.

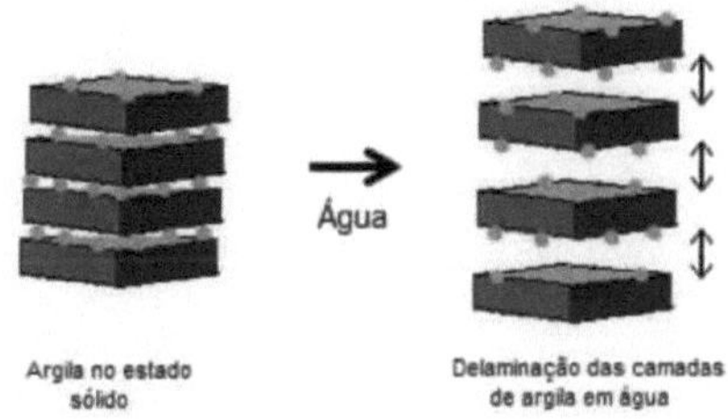

According to Kehew (2006), hydration is a complex result of changes in the water-clay formation system which disturb the internal balance of these particles. The same author indicates that the clay particles are like platelets which have negative charges on their surface and positively charged edges, and so the water molecules are attracted to the surface of the platelet due to the polar nature of the water molecule.

According to Komine (2010), the type, size and charge of exchangeable cations present in the intermediate layer have a major impact on the magnitude of the swelling that occurs in reactive formations. According to the author, interlayer hydration, which is associated with the hydration of exchangeable cations, includes water activity and the size and morphology of the clay particles. In their studies, Hendricks *et al.* (1940) proposed that the expansion process is initiated by the hydration of exchangeable cations.

According to Cardoso (2005), one way of preventing clays from hydrating is to add chemical products called inhibitors. The author mentions that there are many types of hydration inhibitors, which can be inorganic or organic. The mechanism of action of inhibitors basically consists of fixing their cationic fraction on the negative surface of the clay particles (Cardoso, 2005).

Qu *et al.* (2009), investigated the inhibitory property of polyoxyalkylene amine (POAM) according to conventional methods for evaluating laboratory-prepared drilling fluids against sodium montmorillonite (Na-MMT). The study showed that POAM was completely soluble in water and exhibited superior performance in inhibiting the hydration of Na-MMT and significantly reduced the hydration of these formations and shales, indicating good hydration control by this inhibitor.

A study by Zhong *et al.* (2011) evaluated the hydration of reactive shale using the inhibitor polyester diamine (PEDA) in a drilling fluid system. The results indicated that the inhibition properties of PEDA are superior to potassium chloride, which is the conventional type of inhibitor, and can be improved by lowering the pH value. This is possible, according to the authors, due to the low molecular weight of PEDA, which makes it possible for it to be intercalated into the clay structure.

For Wang *et al.* (2011) the inhibition of hydration and dispersion of reactive formations by aqueous fluids must be achieved with inhibitors, and these must maintain the appropriate flow properties of the drilling fluids to play a significant role in the successful drilling process. The authors show that low molecular weight polyether amine provides sufficient inhibition performance and is an excellent additive as it has no effect on the rheological properties of clay dispersions.

2.3 Mechanisms of instability of reactive formations

Xuan *et al.* (2013) reported that accidents caused by the instability of reactive formations in the face of aqueous fluids result in an economic loss of 800 million dollars a year. The same study mentions that, in general, shales are hydrophilic, and this makes it difficult to obtain a drilling fluid that avoids instability problems and meets the general properties required for good drilling fluid performance.

According to Frydman and Fontoura (2001), the behavior of shale during drilling is defined by a complex combination of mechanical, chemical, thermal and electrical processes and its behavior cannot be understood if the mechanisms are analyzed separately. According to the authors, the presence of drilling fluid can lead to the transport of ions and/or water into or out of the formation.

According to Van Oort (2003), there are two controlling mechanisms that explain these phenomena: the first is called hydraulic diffusion, which represents the flow of fluid in response to the hydraulic pressure gradient between the pressure exerted by the drilling fluid and the pore pressure of the formation, and the second mechanism is chemical diffusion due to the difference in chemical potential between the drilling fluid and the pore fluid of the shale. In response, solute migrates from zones of high concentration to areas of lower concentration. Both processes are direct flow mechanisms, since the fluid and solute flows are directly related to their respective gradients.

According to Hawkes *et al.* (2000), the difference in chemical potential between the drilling fluid and the pore fluids of the shales is caused by the concentration gradient and type (nature) of the salts present in the interstitial waters of the shale and in the drilling fluid. The authors also mention that in addition to the osmotic transport of the solvent, the transport of ions can also occur, and this transport is called diffusion.

Niu *et al.* (2013) define osmosis as the movement of water through a semi-permeable membrane that promotes the separation of solutions with different salt concentrations or different chemical activities. This mechanism causes the solvent to migrate from a medium with a lower concentration to one with a higher ionic concentration. The same authors define diffusion as the process by which dissolved ions move from areas of higher concentration to areas of lower concentration through a non-ideal semi-permeable membrane, i.e. osmosis and diffusion occur in opposite directions.

According to Garcia and Fontoura (2007), the chemical activity concert has been applied to oil engineering to quantify the differences in chemical potential between the shale and the drilling fluid. According to Yan and Deng (2013), chemical activity makes it possible to compare the partial molar free energy between two media, and is related to the free energy of the water molecules in a solution. Solutions with a high solute concentration have low water activity and solutions with low concentrations have high water activity.

If the activity of the drilling fluid is greater than the activity of the shale, water flows from the drilling fluid into the formation. If the opposite occurs, i.e. if the activity of the shale is greater than the activity of the drilling fluid, water flows from the formation into the drilling fluid. When the activities on each side of the membrane are equal, the system remains in equilibrium, i.e. no flow occurs (Hawkes *et al.*, 2000), as can be seen in Figure 6.

Figura 6 Water is transported through a perfect semi-permeable membrane at the interface between shale and drilling fluid (Hawkes *et al.*, 2000).

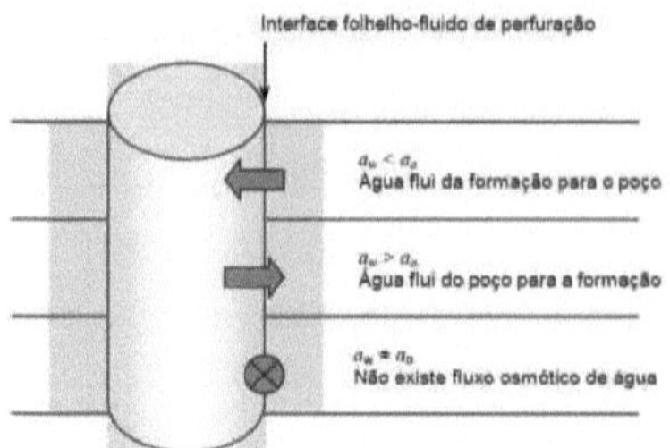

Thus, considering only the activity and no diffusive process, it could be inferred that the use of drilling fluids with a higher saline concentration (low water activities) than the pore fluid of the shales, generates an osmotic potential in the formation which is used to generate (osmotic) flow towards the drilled well, removing water from the formation, causing a drop in pore pressure, increasing the effective tension and consequently improving the resistance of the formation.

When two fluids of different concentrations are separated by a perfect semi-permeable membrane, the water present in the less concentrated solution flows into the region of higher concentration in an attempt to balance the system, thus generating a pressure gradient (OLSEN *et al.*, 1990, GHASSEMI *et al.*, 2009).

Shuixiagn *et al.* (2011) carried out a study with various environmentally friendly aqueous drilling fluids and concluded that the prolonged direct contact between the aqueous phase of the drilling fluid and the shale allows water to be transferred to the latter, causing it to swell. The authors also concluded that this occurs mainly when the chemical potential of the drilling fluid is higher than that of the drilled formation.

According to an article by Al-Bazali (2012), it is possible to manipulate the chemical potential (water activity) of the drilling fluid to mitigate the problems caused by the instability of the well containing shales, i.e. to use the chemical potential as a driving force to extract water from the shales.

Several studies have shown that the instability of shales is due to the flow of water into them, especially when the chemical potential of the drilling fluid is lower than that of the formation, which is why the reduction of water activity by using the addition of salts to promote inhibition (ZHANG *et al.*, 2008; KEIJZER *et al.*, 2004; MODY *et al.*, 2002; SCHLEMMER *et al.*, 2002; EWY AND STANKOVICH, 200 and TAN *et al.*, 1996). The instability caused by the lack of control over the hydration of the shales can be seen in Figure 7.

Figura 7   - Well instability caused by the hydration of shales.

Figure adapted (Last and Plumb, 1995).

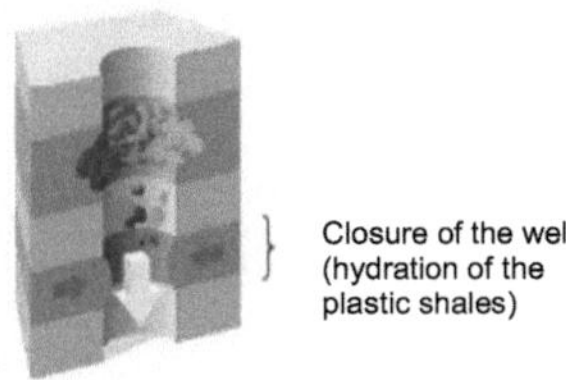

Studies carried out by Suter (2009), as well as Cygan (2009), indicate that research is mostly focused on clays from the smectite group, due to their high swelling potential and the frequency with which this formation is encountered during oil well drilling operations. The authors indicate that the tendency of these formations to swell macroscopically is the main cause of shale instability and can potentially lead to well collapse.

Mokni *et al.* (2010), studied the effect of osmotic phenomena on the swelling of reactive formations, and concluded that the swelling induced by the presence of water can be reduced by dissolving salt crystals and a formulation containing NaCl is proposed to combat the swelling observed in reactive formations.

Bailey *et al.* (2004) applied the dispersibility test to measure the swelling inhibition performance of polyacrylamides with potassium ions in relation to smectite and concluded that inhibition using this ion was an effective procedure.

Diaz-Pérez *et al.* (2007) concluded in their studies that thermogravimetric analysis (ATG) can reveal the distribution of water in a clay in various specific places in the mineral structure and this test is an important tool that can be used to understand the hydration of the swelling process of clay minerals.

Norrish (1963) used X-ray diffraction to measure the spacing of smectite when immersed in saturated salt solutions. A slight increase in the spacing of the clay layers was observed with decreasing salt concentration, thus concluding that there is an increase in the adsorption of water molecules on the clay with decreasing salt concentration in the fluids used.

2.4 Drilling fluids

2.4.1 Definition and functions

Drilling fluids, also known as drilling mud, can be conceptualized according to Amorim (2006) as compositions that are often liquid and intended to aid the process of drilling oil wells.

According to the *American Petroleum Institute* - API, drilling fluids are defined as circulating fluids used in rotary drilling to perform the functions required during the drilling operation. According to Machado (2002), from a physical point of view, drilling fluids behave like non-Newtonian fluids, i.e. the relationship between the shear rate and the strain rate is not constant.

According to Nascimento *et al.* (2010b), the composition of a fluid is selected in order to obtain the ideal properties, such as viscosity, gel consistency, filtrate control, inhibition of hydratable clays and lubricity coefficient, and according to the authors, in today's well-drilling scenarios, environmental requirements are becoming increasingly stringent, thus being a determining factor in the choice of a drilling fluid.

According to Leal *et al.* (2013), the drilling fluid is designed to prevent accidents by counterbalancing the natural pressure of the rock formations. An appropriate balance is achieved and the pressure against the walls of the well is sufficient to counterbalance the pressure exerted by the rock formations, but this must not be too high so as not to damage the well.

Darley and Gray (1988) mention that drilling fluids have been used for over a hundred years and are used in the oil extraction industry, both *in onshore* and *offshore* drilling, in artesian well drilling, as well as in drilling operations.

According to Bourgoyne *et al.* (1991), the use of drilling fluids began in 1901 in the *Spindletop* well in Texas and its development was a major challenge for the oil industry, in the search for the optimum point between cost, technical performance, and from the 1980s onwards, in compliance with environmental requirements. The efficiency of drilling a well depends largely on the relationship between the drilling fluid used and the formations drilled.

Amorim (2006) points out that transporting and removing the fragments and debris generated during drilling are some of the most important functions of drilling fluids. The fluid, when injected into the well, exerts a jetting action that keeps the bottom of the well and the drill bit free of debris, ensuring a longer life for the drill bit and greater drilling efficiency. The efficiency of muds in the gravel removal process depends on several factors: annular velocity, density and viscosity of the fluid.

According to Ferreira *et al.* (2008), drilling fluids in relation to the liquid dispersant medium can be of two types: water-based fluids and non-aqueous fluids. Water-based fluids basically use water and clay from the smectite group, as well as fluids with polymer additives. Still according to the authors, in drilling sensitive to contact with water or with particular technological needs, non-aqueous fluids with organic media are used instead of water as a dispersing agent or aqueous fluids with the use of expansion inhibitors.

2.4.2 Expandable clay and shale inhibitors

The composition of the fluid is variable and depends on the particular requirements of each drilling operation. As the fluids pass through different types of formations, there is a need to use this or that additive that makes up the different types of drilling fluids.

According to Thomas (2001), the ratio between the basic components and the interactions between them cause sensitive changes in the physical and chemical properties of the fluid, and therefore the composition is the main factor to consider when controlling the properties of the drilling fluid.

According to Albano (2011), there are chemical additives, both organic and inorganic, which minimize clay-water interactions. These inhibitors act on the clay's polar ionic active sites located in the basal spacing and on the lateral edges, making it difficult for water molecules to enter and effectively reducing the hydration of the clays (SERRA E SANTOS, 2004; BASSI *et al.*, 2009). The main organic additives used as clay swelling inhibitors are cationic molecules containing quaternary ammonium groups in their structures, however, they are considered expensive by the oil industry.

For Komine (2010), an acceptable clay swelling inhibitor must not only significantly reduce clay hydration, but must also meet increasingly stringent environmental guidelines and be low cost. The development of these inhibitors, which are generally based on water-soluble polymers, therefore represents a challenge for the oil field.

Silva and Almeida (2010) report that for many decades various chemical additives have been used to inhibit the reactivity of shales. The most commonly used precursor fluids were formulated from saline solutions in high concentration, mainly using potassium chloride (KCl) and sodium chloride (NaCl). However, these salts in large quantities negatively affected biological and chemical systems, imposing limitations on their use. As a result, the use of polymer/KCl-based systems has become commonplace.

According to Santos *et al.* (2012), from the mid-1990s onwards, drilling fluids formulated with silicates began to be used mainly in combination with potassium chloride. The level of inhibition of the reactivity of shales using high concentrations of silicates was positive, however, problems related to increased torque in the drill string and the precipitation of silica in the fluids limited the application of fluids containing silicates.

To inhibit clay swelling, the most commonly used additives are inorganic and monovalent cations, such as KCl and NaCl, and bivalent cations, such as $CaCl_2$. Quaternary ammonium salts are also used to stabilize clays.

These ammonium salts are considered to be expensive by the oil industry to be used in drilling fluids, so they are generally associated with sodium salts such as NaCl and potassium salts such as KCl, which are cheaper (Vidal *et al.*, 2007). The cations in these salts are smaller in diameter than water and tend to stay between the layers of clay, thus preventing it from hydrating (SILVA E ALMEIDA, 2010).

Patel *et al.* (1995) studied inhibitors with ammonium ions, which have a similar hydration radius to potassium and act in a similar way. They are also economically attractive. However, they found that the application of ammonium-based inhibitors is limited mainly when subjected to temperatures above 66^0 C (150^0 F) and in environments with very high pH, because in these cases the ammonium salts dissociate into ammonia, creating a dangerous environment for rig workers.

According to Churchman (2002), cationic polymers are widely used in the oil industry, mainly because they are very effective at inhibiting the reactivity of shales, as they interact strongly with the negatively charged layers of reactive shales. However, Rosa *et al.* (2005) warn that the toxicity and incompatibility of cationic polymers with other anionic additives limits the use of the former in drilling fluids with a greater margin of success.

According to Silva and Almeida (2010), the cation acts as a bridge between the layers of clay. Also according to the authors, divalent cations have a greater force of attraction, thus presenting the ability to bind two layers with greater intensity, considerably minimizing swelling. The potassium ion is an exception among the monovalent ions. Due to its small size, it tends to fit perfectly into the space between the layers, unlike the other monovalent cations which hydrate excessively, causing the clay to swell. In general, ions with a smaller hydrated diameter tend to have greater reactivity-inhibiting activity. For this reason, potassium salts are considered an excellent alternative for controlling the expansion of active formations.

Some studies indicate that most inhibitors of reactive formations act, as already mentioned, on the principle of replacing a cationic species in the layer of the expandable formation with a potassium ion, which motivates researchers to use different potassium-based compounds to replace the traditional inhibitors used for this purpose in order to produce better expansion control results (GHOLIZADEH- DOONECHALY *et al.*, 2009; HORTON *et al.*, 2012).

2.4.2.1 Potassium salts

Potassium has a great affinity with polymeric fluids. It has a great inhibiting capacity, preventing the hydration of smectites (hydratable clay) which, with the presence of the ion K^+ in the space between the layers, are transformed into illites (non-hydratable clay). Silva (2005) mentions that smectites are selective (affinity) for potassium, as it is present in sufficient quantity to prevent an increase in the space between the layers of the active formations. In addition, potassium salts have been used as swelling inhibitors in water-based fluids. For Simpson *et al.* (2000), the inhibition is explained by the possible penetration of unhydrated ions into the pores of the clay formation.

A cation can act as a bond or bridge to hold clay mineral particles together or to limit the distance between them. In general, cations have a greater tendency to bind particles together. The potassium ion is an exception because its ionic radius allows it to adapt to the hexagonal lattice space of the oxygen atom on the surface of smectitic clay minerals and because its coordination number is favorable to this arrangement (EWY AND STANKOVICH, 2002).

Al Bazali *et al.* (2011) believe that the transfer of potassium ions to the layers of reactive formations can decrease the distance between the clay plates, which leads to the shrinkage of the swelling, thus improving the stability of the formation.

According to Karaborni *et al.* (1996), the effectiveness of K ions$^+$ in minimizing swelling pressures in montmorillonites is related to the small degree of hydration of these ions in water, resulting in large amounts of ionic repulsion, thus preventing expansion.

In view of this, the application of potassium-rich salts or products has become common practice in the stabilization of boreholes in hydratable claystones, mainly for oil and mining wells (HORSROUD, 1994).

2.4.2.1.1 Potassium chloride

Van Oort *et al.* (2003) indicate in their studies that potassium chloride (KCl) is probably the best known inhibitor in the oil industry. According to the authors, its popularity stems above all from its ability to reduce the swelling susceptibility of smectite clays. It has therefore been applied very effectively in the drilling of shales which generally contain large quantities of these clays (LONG *et al.*, 1976).

According to Al-Bazali and Talal (2012), the main flaw in KCl's performance is its inability to prevent the filtrate from invading the shales. The viscosities of KCl solutions are close to those of water, even at high salt concentrations. It can therefore be seen that the control of filtrate invading the formation cannot be avoided by increasing the concentration of KCl in the fluid.

According to Chesser (1987), salts such as potassium chloride can inhibit clay swelling, however, it is highly flocculent and its presence in the fluid can negatively affect rheology and filtration. Zhong (2011) found that a large amount of potassium chloride is required for effective inhibition, which is harmful to the environment and results in high disposal costs.

2.4.2.1.2 Potassium sulphate

According to a description by Carter *et al.* (2007), potassium sulphate (K_2SO_4) is a non-flammable white crystalline salt that is soluble in water. Its use as a clay expansion inhibitor is due to the supply of potassium ion. The molecular geometry of this salt can be seen in Figure 8.

Figure 8 - Molecular geometry of potassium sulphate.

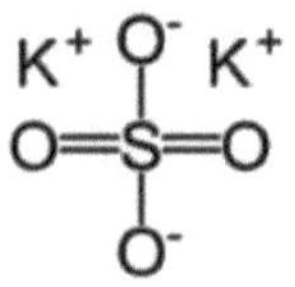

$$K^+ \quad O^- \quad K^+$$
$$O=S=O$$
$$O^-$$

This salt can also be described as an inorganic salt, free of chlorides, which inhibits the expansion of clays due to hydration when drilling in fresh or salt water. It reduces drill bit waxing and improves the drilling rate in plastic clays (PEREIRA, 2007).

According to Baltar and Da Luz (2003), when the well passes through a layer of reactive material, the contact of the water-based fluid with this material causes it to hydrate and expand, which can trap the well drilling tool. Also according to the authors, additives such as potassium sulphate are used to ensure the stabilization of the well and are used in situations where the well passes through layers of shale. Potassium sulphate covers the surface of the shale, preventing it from absorbing water and expanding, obstructing the well.

According to a report carried out by the Oil Research Laboratory of the Federal University of Rio Grande do Norte, polymeric fluids containing potassium sulphate as an expansion inhibitor showed rheological and pH values comparable to fluids without the presence of the inhibitor, which suggests that potassium sulphate does not alter the viscosity or pH values of the fluids. The report also points out that lower lubricity coefficients are obtained for fluids prepared with the inhibitor potassium sulphate, which indicates that the presence of the inhibitor promotes an increase in the degree of lubricity of the fluid.

Jones and Mullin (1974) indicated in their studies that formulations with the presence of potassium sulphate as a swelling inhibitor are indicated as a low cost/benefit ratio alternative, as it is a low cost raw material compared to those normally used by the oil industry.

Potassium sulphate has numerous advantages when used as a swelling inhibitor for reactive formations, the main ones that can be mentioned are (PEREIRA, 2007):

- dissolves easily;

- effectively inhibits the incorporation of clays into the fluid and the hydration of expansive formations;

- provides calibrated holes;

- prevents collapses and entrapment of tools;

- increases penetration speed;

- improves the recovery of samples;

- is fully compatible with a wide range of viscosifiers and fluid compositions,

- is free of chloride ions and environmentally friendly.

2.4.2.1.3 Potassium acetate

Potassium acetate or potassium ethanoate is an ionic compound with the formula $C_2H_3KO_2$ or CH_3COOK. It has a molecular mass of 98.15 g/mol and comes in the form of white crystals or flakes. It melts at 292° C and is soluble in water (256 g/100 mL at 25 oc). According to Cheng *et al.* (2010), potassium acetate is a highly hygroscopic salt obtained by neutralizing acetic acid with potassium hydroxide or potassium carbonate and has the molecular geometry shown in Figure 9.

Figure 9 - Molecular geometry of potassium acetate.

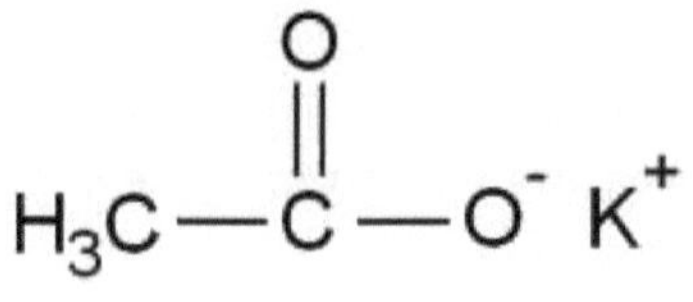

According to studies carried out by Souza and Borges (2011), the use of potassium acetate to control the expansion of shales has some advantages over the use of KCl, among them: acetate solutions have lower water activity than concentrated KCl solutions, giving rise to higher osmotic pressures (providing the osmotic gradient for the dehydration of shales). They can therefore be considered more suitable for reducing filtrate invasion in shales.

According to Van Oort *et al.* (1996), this monovalent salt seems to be particularly suitable for drilling reactive formations, as it reduces the swelling pressure and water content in these formations at the same time. This statement is supported, according to Howard (1995), by field experience. It is important to note that, according to the authors, the benefits mentioned above are generally obtained for highly concentrated saline solutions.

2.4.2.1.4 Potassium citrate

Potassium citrate or tripotassium citrate is a chemical compound with the formula $C_6H_5K_3O_7$. It is the fully neutralized potassium derivative of citric acid. According to Gang-Lu *et al.* (2012), potassium citrate is a white crystalline solid, usually in powder form, with stable physical properties and good solubility in water. It is odorless, has a saline taste and is soluble in water. The density and viscosity of solutions containing potassium citrate increase with increasing concentration. The molecular geometry of potassium citrate is shown in Figure 10.

Figure 10 - Molecular geometry of potassium citrate.

Figure adapted (Melo Jr. *et al.*, 2012).

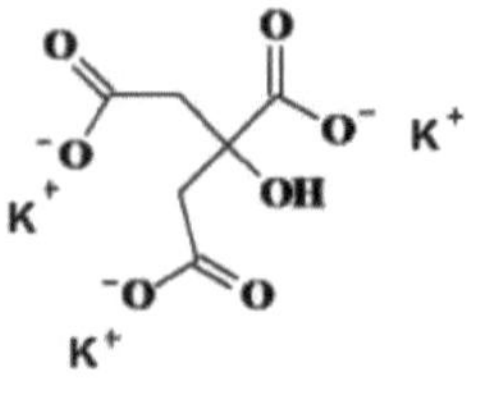

From the above, it can be seen that the problem of controlling the stability of reactive formations can be avoided by using suitable fluids, i.e. those containing inhibitors that prevent the hydration of such formations. In this way, the proposal to develop an inhibited, aqueous, chlorine-free drilling fluid is a work of great scientific contribution and aims to develop an alternative to remedy the problems caused by the hydration of reactive shales.

29

Chapter 3

3. METHODOLOGY

3.1 Materials

3.1.1 Samples studied

Two samples of bentonite clay were studied: a sample of industrialized clay, known commercially as Brasgel PA (the clay was supplied by the company Bentonit Uniao Nordeste Ltda - BUN, located at Avenida Assis Chateaubriand, 3877, Campina Grande, PB), and a sample of imported industrialized bentonite clay, naturally sodic, from the company *Southern Clay*, Texas, United States, known as Cloisite, both with a high degree of swelling.

Shales from various regions of the country were also studied: the Rio do Peixe Basin - PB, the Araripe Basin - PE, the Reconcavo of Bahia, the Parnaiba Basin, among others. Figures 11 to 15 show the samples used in the study, followed by Table 1, which describes and names all the samples studied.

Figure 11 : Samples F1, F2 and F3.

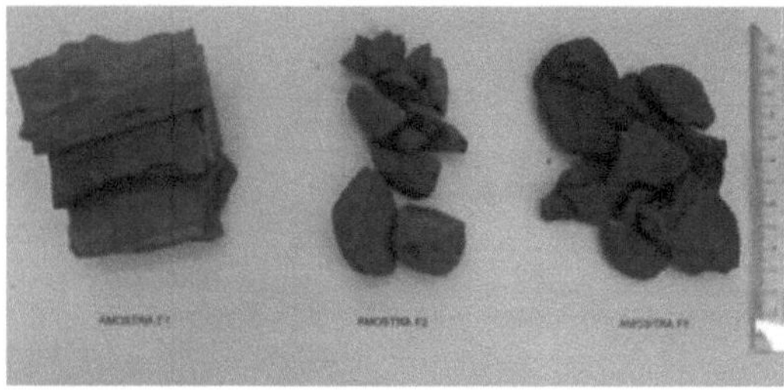

Figure 12: Samples F4, F5 and F6.

Figure 13: Samples F7, F8 and F9.

Figure 14: Samples FIO, F11 and F12.

Figure 15: Samples F13 and the Brasgel PA and Cloisite clays.

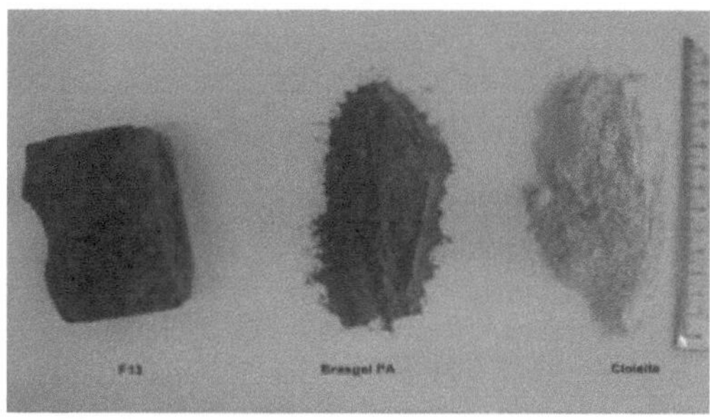

Table 1 - Identification used for the samples studied.

NOMENCLATURE USED	TYPE OF TRAINING	STATE OF ORIGIN	REGION OF ORIGIN
F1	Leaf	Pernambuco	Araripe Basin
F2	Leaf	Bahia	Reconcavo Baiano Basin
F3	Leaf	Bahia	Reconcavo Baiano Basin
F4	Leaf	Pernambuco	Araripe Basin
F5	Leaf	Pernambuco	Cape Basin
F6	Leaf	Bahia	Central Tucano Basin
F7	Leaf	Tocantins	Parnaiba Basin
F8	Leaf	Tocantins	Parnaiba Basin
F9	Leaf	Piaui	Parnaiba Basin
F10	Leaf	Piaui	Parnaiba Basin
F11	Leaf	Piaui	Parnaiba Basin
F12	Leaf	Tocantins	Formation of the Amazon
F13	Leaf	Paraiba	Rio do Peixe Basin
Brasgel PA	Brasgel PA clay	Paraiba	Boa Vista
Cloisite	Cloisite clay	Texas	Gonzales

3.1.2 Additives

The following additives were used to prepare the polymeric drilling fluids: defoamer (silicone-based liquid), viscosifier (xanthan gum), reducer/filter (low viscosity carboxymethylcellulose), pH controller (magnesium oxide), expansion inhibitors (potassium sulphate, potassium acetate, potassium citrate and potassium chloride), the latter being used for comparison purposes as it is widely used by the oil industry), bactericide ((tetrakis)hydroxymethylphosphonium solution)), lubricant (high lubricity vegetable oil chemically treated with acids and alkaline neutralizers), and sealant (calcite).

The basic formulation for preparing the drilling fluids used in this study is described in Table 2. The concentration of the inhibitor was defined on the basis of the results in section 4.2. The base formulation was obtained from Lucena (2012a).

Table 2: Basic formulation of the drilling fluids to be studied.

Additive	Concentration (/ 350mL of water)
Defoamer	0,084g
Viscosifying	1,5g
Filtrate reducer	3,5g
pH controller	1,0g
Expansion inhibitors	20,0g
Bactericidal	0,7g
Lubricant	3.0% of the volume of H2O
Sealant	15,0g

The additive samples were supplied by System Mud Indùstria e Comèrcio Ltda. - acquired in 2012 by Imdex Limited Company located at Rua Otavio Muller, 204, Carvalho, Itajai, SC.

3.2 Methods

The following steps were used to study and evaluate the efficiency of the fluids proposed in the research to control the hydration of reactive formations, according to the flowchart shown in Figure 16.

Figure 16: Flowchart of the steps for studying the efficiency of the proposed fluids in controlling the hydration of reactive formations.

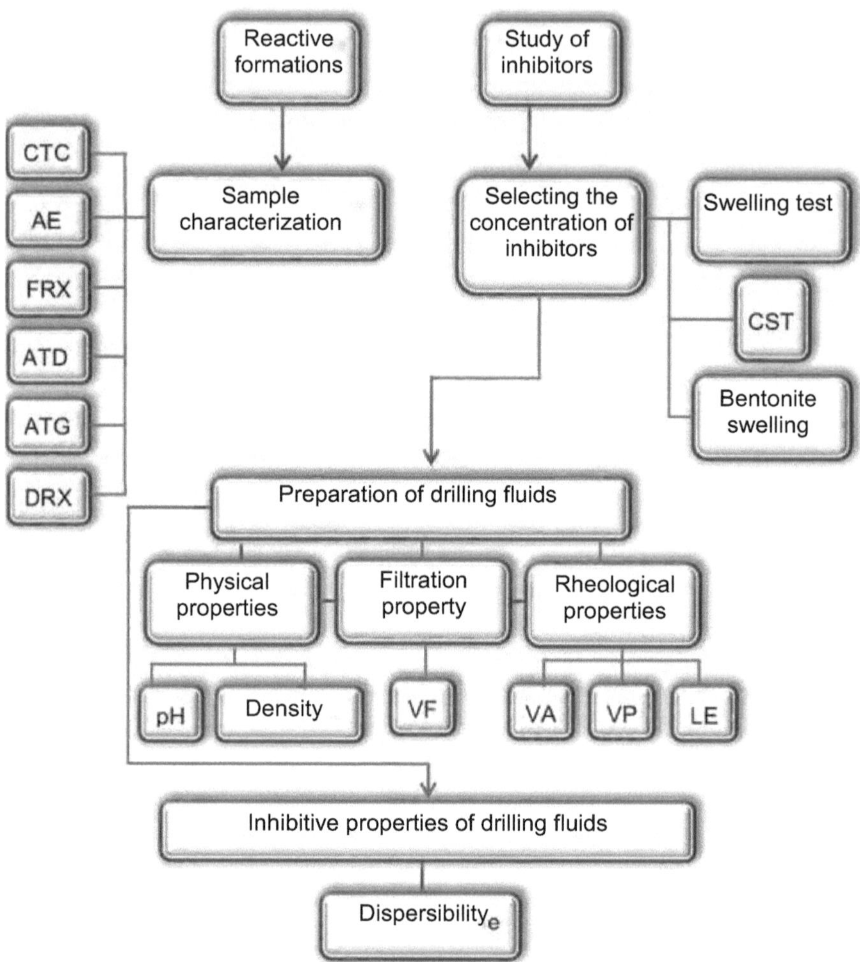

3.2.1 Characterization of shales

The characterization of shales and clays aims to help predict instabilities generated by interactions between drilling fluids and the study samples, which can occur when oil well drilling operations pass through this type of formation.

3.2.1.1 Cation exchange capacity using the methylene blue adsorption method

The cation exchange capacity (CEC) is a property that derives from the structural characteristics of a clay mineral and the determination of this parameter is indispensable for a more accurate evaluation of the samples, since each clay mineral has a range of characteristic values.

The cation exchange capacity (CEC) of the clay was determined using the methylene blue adsorption method (Ferreira *et al.*, 1972). The principle of the test is to introduce increasing amounts of methylene blue solution, in successive doses, until the surface of the particles, which have adsorption capacity, is covered. At this point, there is an excess of methylene blue in the preparation, corresponding to the turning point which marks the end of the test, and which can be detected by the spot test.

For this method, a suspension of 0.1g of reactive formation in 20mL of water was prepared. To the suspension was added 1.0mL of methylene blue with a concentration of 4.31×10^{-3} mol L^{-1} . With each addition of methylene blue, the suspension was homogenized for 2 minutes and, using a glass rod, a drop of it was placed on a sheet of Whatman filter paper n° 50. This procedure was repeated until a light blue color appeared around the circle formed by the adsorbed drop. The persistence of the bluish ring indicates that the saturation point has been reached. The presence of a halo indicates the existence of free methylene blue in the preparation, showing that all the particles susceptible to adsorbing methylene blue are covered by a layer of methylene blue molecules. This stage was carried out in the Drilling Fluids Research Laboratory - PEFLAB at LABDES/ UFCG.

3.2.1.2 Specific area

The specific area (SA) of the natural and industrialized clay samples was determined using the methylene blue adsorption methods described in Ferreira *et al.* (1972).

This stage was carried out in the Drilling Fluids Research Laboratory (PEFLAB) at LABDES/ UFCG.

3.2.1.3 Particle size analysis by laser diffraction (AG)

Particle size analysis by laser diffraction uses the method of dispersing particles in a liquid phase associated with an optical measurement process using laser diffraction. In this method, the proportional relationship between laser diffraction and particle concentration and size is combined. To carry out this characterization, the samples of shales and clays were passed through an ABNT sieve n° 200 (0.074mm), and dispersed in 250mL of distilled water in a Hamilton Beach N5000 shaker at a speed of 17,000rpm for 10min, then these dispersions were placed in a CILAS model 1064 equipment, in wet mode, until they reached the ideal concentration, which is 150 diffraction units/area of incidence.

The determination of particle size distribution by laser diffraction was carried out on a CILAS 1064 laser granulometer, available at the UAEMa/CCT/UFCG Materials Characterization Laboratory.

3.2.1.4 X-ray fluorescence

The identification of mineral oxides is one of the main parameters used to estimate the reactivity potential of these rocks and chemical analysis can help with this.

The chemical analysis of the clays by X-ray fluorescence (XRF) was carried out using the semi-quantitative method under a nitrogen atmosphere. The material supplied was quartered and pressed in a manual tablet press, with a diameter of approximately 15mm. The equipment used will be the SHIMADZU EDX-720 model belonging to the Materials Engineering Characterization Laboratory/CCT/UFCG, Campina Grande, PB.

3.2.1.5 Differential thermal analysis

Differential thermal analysis (DTA) determines the difference in temperature of a sample in relation to an inert reference, in this case aluminum oxide (Al_2O_3). Endothermic and exothermic transformations appear as deflections in opposite directions on the thermodifferential curve. The samples were analyzed using a Thermal Analysis system model RB-3000-20, operating at 12.5° C/min. The maximum temperature used in the differential thermal analysis was 1000° C and the standard used in the DTA tests was calcined aluminium oxide (Al_2O_3), as mentioned above.

The tests were carried out at the Ceramics Laboratory of UAEMa/CCT/UFCG, Campina Grande, PB.

3.2.1.6 Thermogravimetric analysis

Thermogravimetric analysis (TGA) determines the change in mass of the sample as a function of temperature and/or time. The samples were analyzed using a Thermal Analysis system model RB-3000-20, with a maximum temperature of 1000 C.°

The tests were carried out at the Ceramics Laboratory of the UAEMa / CCT / UFCG, Campina Grande, PB.

3.2.1.7 X-ray diffraction

The X-ray diffraction (XRD) technique was used to identify the mineralogical constituents of the samples by means of a qualitative study. X-ray diffraction (XRD) is a method widely used in the characterization of crystalline lattice structures, and is therefore a very useful method for the qualitative identification of some of the components in the shales to be studied in this work.

The X-ray diffraction (XRD) analyses of the samples, in dry form and treated with ethylene glycol, were carried out using a SHIMADZU XRD-6000 X-ray diffractometer operating with copper κ-alpha radiation, with a voltage of 40kV and 30mA of current and a wavelength of λ =1.5406A. The samples were analyzed with a scan between 2θ (30) and 2Θ (70o), the speed of the goniometer was 20/minute.

The tests were carried out at the Materials Engineering Characterization Laboratory / CCT / UFCG, Campina Grande, PB.

3.2.2 Selecting the best concentrations of clay swelling inhibitors

In order to select the best concentrations of inhibitors, swelling tests, determination of free water by capillary suction and the bentonite inhibition test were carried out.

3.2.2.1 Swelling test

In order to select the concentrations of the inhibitors that give the best results in terms of their performance and to assess the degree of inhibition of expansive clays by the chemical inhibitors, the swelling test was carried out. The test is based on the *Standard Test Method for Swell Index of Clay Mineral Component of Geosynthetic Clay Liners* (ASTMD 5890-11).

To do this, in a 100mL beaker containing 90mL of water and water solutions with inhibitors (potassium sulphate, potassium acetate, potassium citrate and potassium chloride) at the defined concentrations (16, 18 and 20g of inhibitor/350mL of water), 1g of dry bentonite clay passed through an ABNT sieve nº 200 (0.074mm) was slowly added. The solution was left to stand for 10 minutes. Increments of 0.1g of clay were then added to the solution every 10 minutes until a total mass of 2g of clay was reached. After the entire mass of clay had been added, the sides of the beaker were carefully washed to remove any particles that had adhered to the container. This was done until the volume of the beaker had reached 100mL. The solutions were left to stand for 16 hours and after this time the swelling recorded for each sample was read. The same procedure was carried out for Brasgel PA and Cloisite clays. This stage was carried out in the Drilling Fluids Research Laboratory - PEFLAB at LABDES/ UFCG.

3.2.2.2 Determination of free water by capillary suction

FANN's *Capillary Suction Timer*, model 44.000, is a device that measures the time it takes for a given amount of free water to travel through a cellulose plate using detection electrodes. According to Vidal (2007) this method is used to indicate the permeability of the plaster formed by a water-based drilling fluid, and the amount of free water when in the presence of activated clay. The shorter the suction time, the greater the amount of free water and, consequently, the lower the water-clay interaction. According to Vidal (2006) the advantage of its use is the simplicity of the operation.

The measurement is made by placing 5mL of a dispersion (the dispersion contains water, inhibitor and clay) inside a cylinder in contact with filter paper of a specified thickness. The electrodes are positioned 0.5 and 1cm from the edge of the cylinder and connected to a timer, allowing the time required for the filtrate to flow rapidly through a 0.5cm radius to be measured. The dispersions used were prepared 24 hours before the test.

The results of the test were evaluated after reading the values expressed on the device, and plotted on a graph relating the suction time to the increase in concentration. The CST experiments were carried out in the Drilling Fluids Research Laboratory - PEFLAB at LABDES/UFCG.

3.2.2.3 Benthic inhibition test

The bentonite inhibition test is a method that determines the maximum amount of clay that can be inhibited by a solution containing inhibitor at a concentration of 8g/350mL of water. This test is carried out by slowly incorporating clay into an inhibitor solution in order to simulate what happens when drilling in the field for formations with active shales.

The procedure based on the methodology developed by Patel *et al,* (1995) consists of daily incorporation of clay at a concentration of 10g/ 350mL of water into the inhibitor solution described above and obtaining the rheological properties of the mixture after it has been subjected to a temperature of around 66^0 C (150° F) for 16 hours, before adding a new portion of clay at a concentration of 10g/ 350mL of water. The readings were taken on a Fann 35A viscometer at a speed of 3 rpm. This procedure was carried out until the fluid reached a viscosity value that exceeded the equipment's ability to read this property. The results were compared with those obtained from carrying out the same process with the KCl inhibitor.

Based on the results obtained for the Foster swelling, capillary suction time and bentonite swelling tests, the best inhibitor concentration was selected and used to develop the drilling fluids.

3.2.3 Preparation of drilling fluids

Water-based drilling fluids were prepared with the following additives: defoamer, viscosifier, filtrate reducer, pH controller, chlorine-free inhibitors, bactericide, lubricant, sealant and pH controller in a Hamilton Beach mechanical agitator, model 936.

The drilling fluids were prepared according to field practice, which consists of adding the additives, one by one, under agitation at a constant speed of 13,000 rpm in a *Hamilton Beach* model 936 agitator, in the order described above, remaining under agitation for 5 minutes with each addition of additive, with the exception of the viscosifier, filtrate reducer and sealant, which remain under agitation for 10 minutes.

3.2.4 Determining the physical, rheological and filtration properties of drilling fluids

3.2.4.1 pH measurements of drilling fluids

The pH measurements of the drilling fluids were carried out using a Gehaka model PG 1000 digital pH meter.

3.2.4.2 Density measurements

In order to obtain the density of the formulated fluids, density tests were carried out using a Fann model 140 mud scale. Density values are expressed in g/cm .^{3}

3.2.4.3 Rheological properties

The fluids were aged in a Fann model 704 ES Roller Oven at a temperature of 66° C (150° F) for 16 hours.

For the rheological study after aging, the fluid was transferred to a Fann thermal cup at a temperature of 49° C (120ƟF), and the procedure for obtaining the readings on the Fann 35A viscometer was carried out, as shown below.

After resting for 24 hours, a rheological study of the drilling fluids was carried out. To do this, the fluid was agitated for 5 minutes in a Hamilton Beach mechanical agitator, model 936, at a speed of 17,000 rpm. After stirring, the fluid was transferred to the container of the Fann model 35A viscometer. The viscometer was operated at a speed of 600 rpm for 2 minutes and a reading was taken. Then the speed was changed to 300 rpm and the reading was taken after 15 seconds.

The apparent viscosity (VA) is the value obtained from the reading at 600 rpm divided by 2, given in cP, and the plastic viscosity (VP) is the difference between the readings obtained at 600 rpm and 300 rpm, also given in cP. The flow limit (LE) is the difference between the reading obtained at 300 rpm and the plastic viscosity (VP) according to API standard 13 B-1 (2003). The rheological study was carried out at the Drilling Fluids Research Laboratory - PEFLAB at LABDES/UFCG.

3.2.4.4 Determination of API filtrate volumes

The volume of *American Petroleum Institute* - API filtrate (VFAPI) was determined for the aged fluids.

VFAPI was determined in an LPLT (*low pressure low temperature*) filter press, according to the API (2005) standard, applying a pressure of around 7.0kgf/cm^2 (100 psi) for 30 minutes. The results are expressed in mL.

The tests to determine the volume of filtrate were carried out in the Drilling Fluids Research Laboratory (PEFLAB) at LABDES/UFCG.

3.2.5 Determining the inhibitive properties of drilling fluids

3.2.5.1 Dispersibility test

To carry out the dispersibility tests, it is first necessary to prepare the samples for analysis. To do this, the shale samples were ground between ABNT sieves n^0 4 and ABNT n^0 8 (aperture between 4.75 mm and 2.36 mm), which should retain as much as possible of the humidity of the place from which they were extracted, so they should not be dried in an oven or in the air. The equipment used to carry out the dispersibility test was a Fann model 704 ES Roller Oven. This equipment consists of stainless steel cells with a capacity of 400mL.

350mL of drilling fluid was added to each cell of the rotary kiln, followed by 20g of previously prepared shale sample (ground and sieved); this mixture was stirred gently with a spatula to separate the shale particles. The cells were kept rotating at 50 rpm at 1500 F for 16 hours.

After this stage, the cells were cooled to room temperature. The cell contents were then carefully filtered through an ABNT n^0 100 mesh sieve (opening 0.150mm), with a fresh water flow rate of around 2 liters/minute to wash the sample. Material with a particle size of less than 100 mesh was considered dispersed and had to be dried in an oven at 60°C until it reached a constant weight.

The dispersibility value in percentage is an average of two measurements and is calculated using the following formula:

$$D = \left[\frac{(Pi - Pr)}{Pi}\right] x100$$

Where:

D = Dispersibility, %

Pr = Weight of retained shells, g

Pi = Initial weight of the shale sample, g

Chapter 4

4. RESULTS AND DISCUSSIONS

4.1 Characterization of shales and clays

The characterization of the samples studied (shales and clays) was carried out by means of cation exchange capacity and specific area, based on chemical analysis (X-ray fluorescence), differential thermal analysis, thermogravimetric analysis and X-ray diffraction.

4.1.1 Cation exchange capacity and specific area

Tables 3 and 4 show, respectively, the typical CTC values of some clay minerals and the CTC values obtained for the samples listed above in Table 1.

Table 3- Characteristic CEC values of important clay minerals (Sousa Santos, 1992).

Argillomineral	CTC (meq/ 100g)
Kaolinite	3-15
Ilita	10-40
Vermiculite	100-150
Smectite	80-150

Table 4- Cation exchange capacity (CEC) for the samples studied.

Sample name	CTC (meq/ 100g)
F1	72
F2	66
F3	52
F4	48
F5	44
F6	12
F7	24
F8	36
F9	32
F10	34
F11	32
F12	26
F13	28
Brasgel PA	84
Cloisite	92

According to Sousa Santos (1992), the smectite content of a reactive formation to be drilled is often the only parameter for choosing the right drilling fluid, due to the high expansibility potential of this clay mineral (which can be seen in the CTC range presented). It is therefore possible to correlate the CTC of the samples studied with their reactivity potential.

Rabe (2003) classifies shales with CTC between 47 - 49 meq/ 100g as shales with high cation exchange capacity values. Thus, comparing the results shown in Table 4, it can be seen that around half of the samples analyzed show some degree of reactivity.

From the data presented in Table 4, it can be seen that samples F1, F2, Brasgel PA and Cloisite (72, 66, 84 and 92 meq/ 100g, respectively) have higher cation exchange capacities, i.e. they are more likely to absorb water and consequently have higher expansion rates. Comparing these results with Table 3, it can be said that the Brasgel PA and Cloisite samples have CTC values within the range established for smectite (80 - 100 meq/ 100g), thus indicating that these samples are probably composed essentially of this type of clay mineral, which probably explains their tendency to hydrate.

Samples F3, F4 and F5 have considerable CTC values based on the values obtained by Rabe (2003) for the CTC of reactive shales, which may be an indication of the presence of smectite-type clay minerals in their general composition, however, however, the content of this clay mineral is probably lower than in the previously mentioned samples with higher CTC values. The swelling tests carried out on the samples with significant CTC values showed moderate swelling, which reinforces the tendency for these shales to expand.

Moderate to low CTC values were observed in the other shales studied (F6, F7, F8, F9, F10, F11, F12 and F13). These consequently show little or no reactivity, which results in no or low hydration. It can therefore be said that the F1, F2, F3, F4 and F5 shales and the Brasgel PA and Cloisite clays are the samples with moderate to high reactivity according to the classifications of Souza Santos (1992) and Rabe (2003).

The swelling tests carried out on the shale samples in the presence of water confirmed this trend, i.e. shales F6, F7, F8, F9, F10, F11, F12 and F13, in which the lowest CTC values were found, also showed zero swelling values, making it possible to correlate the low cation exchange capacity with a low tendency to hydrate.

Tables 5 and 6 show, respectively, the specific area values for important clay minerals according to the classification made by Souza Santos (1992) and the specific area results for the samples studied.

Analyzing the AE values in Table 5 and Table 6 together, we see a range from 93.65 m^2 / g (sample F6) to 717.99 m^2 / g (Cloisite clay). It can be seen that the AE values of samples

F1, F2, F3, F4, F5, Brasgel PA and Cloisite were within the typical AE values of montmorillonitic clays and were therefore those with the highest probability of expansion, i.e. probably indicating the greatest capacity to adsorb water, and therefore the highest expansion rates. (SOUZA SANTOS, 1992; AMORIM, 2003; CAMPOS, 2007).

Table 5- AE characteristic values of important clay minerals (Sousa Santos, 1992).

Argillomineral	AE (m^2 / g)
Kaolinite	5-10
Ilita	100-200
Vermiculite	300-500
Smectite	700-800

Table 6- Specific area (SA) for the samples studied.

Sample name	AE (m^2 / g)
F1	561,60
F2	514,80
F3	452,40
F4	374,42
F5	343,39
F6	93,65
F7	187,30
F8	280,95
F9	249,74
F10	265,35
F11	249,74
F12	202,91
F13	218,52
Brasgel PA	655,56
Cloisite	717,99

4.1.2 Particle size analysis

Figures 17, 18 and 19 and Table 7 show the particle size distribution curves of the shale and clay samples studied.

Figure 17- Particle size analysis for samples a) F1, b) F2, c) F3 and d) F4.

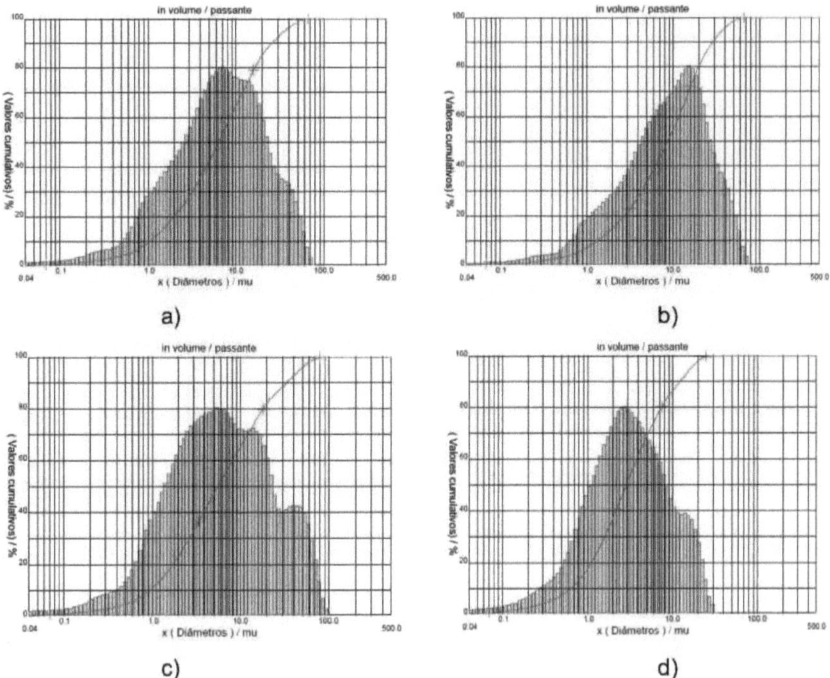
in volume / passante
(Valores cumulativos) / %
x (Diâmetros) / mu
a)
in volume / passante
(Valores cumulativos) / %
x (Diâmetros) / mu
b)
in volume / passante
(Valores cumulativos) / %
x (Diâmetros) / mu
c)
in volume / passante
(Valores cumulativos) / %
x (Diâmetros) / mu
d)

Figure 18- Particle size analysis for samples a) F5, b) F6, c) F7, d) F8 e) F9 and f) FlO.

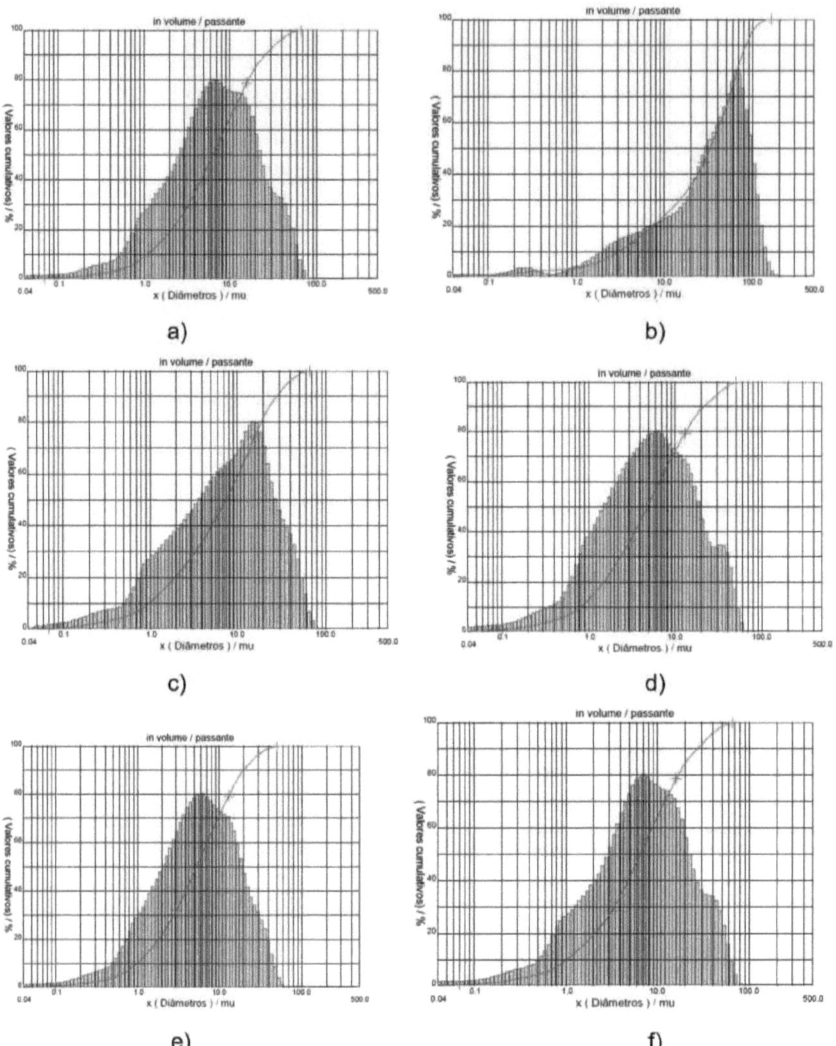

Figure 19- Particle size analysis for samples a) F11, b) F12, c) F13, d) Brasgel PA and e) Cloisite.

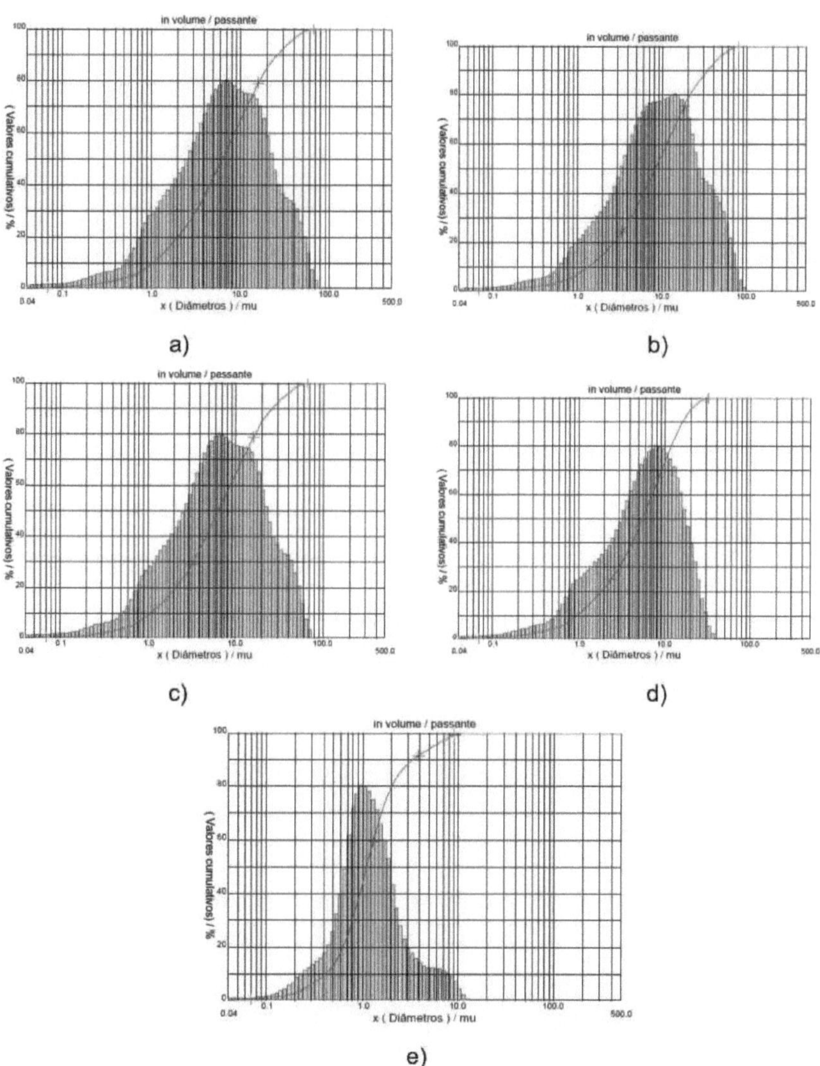

Table 7 - Particle size analysis for the samples studied.

Sample name	Average diameter (µm)	Diameter at 10% (µm)	Diameter at 50% (µm)
F1	8,410	0,82	21,47
F2	8,320	0,99	20,01
F3	11,68	0,95	33,21
F4	4,740	0,65	12,16
F5	8,300	0,95	19,94
F6	38,85	3,05	83,48
F7	11,41	0,96	28,07
F8	10,40	1,05	25,77
F9	10,46	1,09	25,82
F10	10,72	1,03	26,92
F11	12,22	1,10	30,71
F12	12,36	1,35	28,96
F13	13,36	1,36	33,70
Brasgel PA	7,190	0,97	15,92
Cloisite	1,610	0,41	3,32

An analysis of Figure 17 and Table 7 shows a concentration of particles between 0.82 and 21.47µm, with a mean particle diameter of 8.41µm. It was also possible to observe a considerable accumulated volume with a mean diameter of less than 2.0µm, equivalent to 26.06%. Figure 17b shows an average diameter of 8.32µm and a particle concentration of between 0.99 and 20.01µm, as well as an accumulated volume of 21.90% with an average diameter below 2.0µm.

Figure 17c, referring to the particle size analysis of sample F3, shows a distribution of particles between 0.95 and 33.21µm, a mean diameter of 11.68µm, and a considerable accumulated volume with a mean diameter below 2.0µm, equivalent to 23.17%. Figure 17d shows an average particle diameter of 4.74µm, particle distribution between 0.65 and 12.16µm, and a high accumulated volume with an average diameter below 2.0µm equivalent to 37.17%.

Figures 18a, 18b, 18c, 18d, 18e and 18f and Table 7 show average particle sizes of 8.30µm, 38.85µm, 11.41µm, 10.40µm, 10.46µm and 10.72µm, respectively. The samples mentioned above show an accumulated volume with an average diameter below 2.0µm equivalent to 22,94%, 6.57%, 10.44%, 19.85%, 19.11% and 19.36%, in the same order as above, while the particle concentration ranges varied from 0.94 to 19.94µm for sample F5, 3.05 to

83.48µm for sample F6, while sample F7 showed a range between 0.96 and 28.07µm. For sample F8, the particle size concentration ranged from 1.05 to 25.77µm, and intervals of 1.09 and 25.82µm, as well as 1.03 to 26.96µm for samples F9 and F10, respectively.

Figures 19a, 19b, 19c, 19d and 19e together with Table 7 show an average particle diameter of 12.22µm, 12.36µm, 13.36µm, 7.19µm and 1.61µm, respectively, particle concentration in the ranges 1.10 to 30.71µm, 1.35 to 28.96µm, 1,36 to 33.70µm, 0.97 to 15.92µm and 0.41 to 3.32µm, respectively, and average diameter of 12.22µm for sample F11, 14.97µm for sample F12, 13.36µm for sample F13, 7.19µm for the Brasgel PA clay sample and 1.61µm for the Cloisite clay sample.

It can be seen that Cloisite clay had a very small average particle size, which may be related to the purity of the sample and follows the characteristics obtained for the clay mentioned in studies carried out by Leite *et al.* (2008), which indicated low average diameter values for the clay mentioned. Based on these results, it can be concluded that the Cloisite clay has a higher clay fraction than the Brasgel PA clay, which is consistent with the specific area results.

The Brasgel PA sample showed a particle size analysis close to that obtained by Santos and Amorim (2013). For this analysis, the authors observed an average particle diameter of 7.49µm, a percentage of particles below 2µm of 19.79% and a particle size variation between 1.02 and 16.50µm.

According to Souza Santos (1992), clay fractions have particle sizes below 2µm, so a high percentage of particles in the range below this value is indicative of the presence of clay in the composition of the samples studied. This indicates that samples F1, F2, F3, F4 and F5, as well as Brasgel PA and Cloisite clays, have considerable percentages of particle concentration below 2µm, thus indicating considerable clay content in their compositions. This result is in line with that obtained for CTC and AE, i.e. the samples with the highest CTC and AE values also had a considerable concentration of particles below 2µm in size.

4.1.3 X-ray fluorescence

Table 8 shows the results of the chemical compositions obtained for the shales listed in Table 1 using the X-ray fluorescence (XRF) technique.

Table 8- Chemical analysis of the constituents of the shales.

Percentage composition (%)

Samples	SiO2	Al2O3	Fe2O3	K2O	MgO	TiO2	MnO	BaO	SO3	P2O5	CaO	Na2O
F1	57,33	21,25	8,71	5,61	3,98	1,03	0,38	0,30	0,21	0,07	-	0,85
F2	53,46	17,00	8,65	3,94	5,15	0,72	0,13	0,34	0,28	0,05	9,27	0,71
F3	57,67	23,61	7,63	3,98	3,28	0,84	0,46	-	0,74	-	0,46	0,54
F4	52,24	21,11	5,21	2,44	4,38	0,55	0,40	0,20	0,64	-	0,56	0,62
F5	54,91	18,25	6,82	3,71	5,52	1,03	0,07	-	2,73	0,54	3,09	0,75
F6	37,75	12,12	6,46	2,29	-	0,75	-	-	22,95	1,18	12,99	-
F7	60,80	33,18	1,26	1,98	0,97	-	-	0,27	0,25	-	-	-
F8	48,02	29,50	7,46	2,72	-	1,05	-	0,35	9,87	-	-	-
F9	55,56	35,11	2,29	2,56	1,11	1,31	-	0,32	1,65	-	-	-
F10	53,27	31,61	6,10	3,33	1,16	1,28	0,02	-	3,08	-	-	-
F11	50,63	33,37	4,64	3,02	0,93	1,10	-	0,35	5,69	0,20	-	-
F12	56,77	27,14	7,41	3,78	2,01	0,99	0,08	0,31	0,27	0,45	0,68	-
F13	54,14	20,00	9,43	4,17	4,26	0,82	0,13	-	0,19	0,29	6,09	-
Brasgel PA	64,11	18,53	9,39	0,44	2,66	0,88	0,04	0,23	0,28	0,26	1,27	1,84
Cloisite	60,89	21,12	4,80	0,06	2,20	0,10	-	-	-	-	0,20	4,20

Table 8 shows that all the shales studied have a higher proportion of silicates and aluminates, which probably indicates the existence of quartz (sio2) and clay minerals such as kaolinite, smectite and illite in their composition.

According to Celik (2010), an important aspect in relation to the chemical composition is the high quantity of some oxides such as Fe2O3, Na2O, K2O, CaO and MgO, present in some samples, such as F1 and F2. Comparing the result of this percentage sum with the mineralogy indicates that it is associated with the presence of clays with an occurrence of smectite.

Rodrigues *et al.* (2004) state that minerals with appreciable S1O2 content and Al2O3 content above 46% are classified as silica-aluminous. The same authors indicate that considerable levels of MgO are indicative of the presence of the clay mineral smectite.

According to Mota *et al.* (2008), silica or silicon oxide (sio2) is related to clay minerals, feldspars and quartz. A sample with high levels of SiO2 and Al2O3 and appreciable levels of MgO and CaO probably indicates the presence of the clay mineral smectite.

One of the fundamental parameters for understanding the properties of a mineral is the quantification of the main oxides that make it up. Table 9 shows the predominance of silicon oxides in the range of 37.75% to 64.11%, and Al2O3 between 12.12% and 35.11%.

According to Prado *et al.* (2012), these oxides may be mainly associated with the clay minerals present in the samples.

The low content of alkaline oxides and the high concentrations of aluminum and silicon oxides in some samples can be attributed to the presence of kaolinite, which naturally has a low percentage of alkaline oxides. The concentrations of CaO, MgO, Na_2O and K_2O ranged from 0.45 to 12.99%, 0.93 to 5.52%, 0.54 to 4.20% and 0.06 to 5.61% respectively for the samples studied. It can also be seen that most of the samples analyzed had a considerable iron oxide content in their compositions.

F1 has around 57% SiO_2, which, according to what has already been mentioned, may indicate the presence of quartz in its composition. The presence of Al_2O_3 is also significant (21.25%). The presence of SiO_2 together with Al_2O_3 indicates the presence of clay minerals such as kaolinite, smectite and illite, which may explain the certain reactivity of the shale according to the CTC results. The K_2O content of 5.61 may indicate the presence of illite.

Based on a sample of reactive clay studied by Souza Santos (1992) when he discovered the clays of Boa Vista, PB_1 , the following composition was found: 51.10% SiO_2, 17.30% Al_2O_3, 6.78% Fe_2O_3 and 0.55% K_2O. It can be seen that some of the percentages obtained for the reactive clay are similar to the F1 shale (SiO_2, Al_2O_3 and Fe_2O_3), but it has a higher percentage of K_2O, which may indicate the stronger presence of illite in its composition.

When analyzing sample F2, it can be seen that, like the F1 rock, it has a high SiO_2 content (53.46%). There is also a considerable presence of Al_2O_3 (18.00%), a value which is very close to that obtained by Souza Santos (1992) cited above. There is also a considerable presence of Fe_2O_3 (8.65%) and K_2O (3.94%), indicating the presence of illite. The presence of calcium and magnesium may highlight the polycationic nature of the sample.

Like samples F1 and F2, sample F3 has a high content of SiO_2 (57.67%) and Al_2O_3 (24.61%). The presence of Fe_2O_3 in the 6 to 9% range indicates the presence of illite in its composition. A highlight of the F3 analysis was the 3.98% MgO content, which according to Rodrigues *et al.* (2004) and Mota *et al.* (2008) is indicative of the presence of the clay mineral smectite.

An analysis of sample F4 shows significant levels of SiO_2 (52.24%) and Al_2O_3 (22.31%), which, as already mentioned, may be indicative of the presence of quartz and clay minerals. Sample F5 has a high content of SiO_2 (54.91%) and Al_2O_3 (18.25%). The presence of Fe_2O_3 in the 6 to 9% range indicates the presence of illite in its composition. There is a considerable content of MgO (5.52%), which, as mentioned above, is correlated with the presence of the clay mineral smectite.

From what has been discussed in relation to the previous samples, we can detect indications of the presence of clay minerals such as illite and smectite, as well as indications of the presence of quartz. The presence of Fe_2O_3 and K_2O may reinforce the presence of illite in the sample, and the considerable percentage of MgO (5.52%) may be indicative of the presence of carbonates.

Sample F6 has a SiO_2 content (37.75%) that is relatively lower than the levels obtained for the other samples studied, although the presence of this compound indicates the presence of quartz.

Analysis of sample F7 shows a high percentage of SiO_2, while the presence of K_2O (1.98%) may indicate the presence of illite.

Sample F8 shows a high SiO_2 content (48.02%). The presence of Fe_2O_3 may indicate the presence of illite, since it has a content of (7.46%). The presence of illite is also evidenced by the presence of the alkaline oxide K_2O (2.72 %).

The analysis of sample F9 shows that the high SiO_2 content (55.56%) and low Fe_2O_3 content (2.29%) may indicate a low content of smectite-type clay minerals. There is also a presence of K_2O (2.56%) and a significant presence of SO_3 (1.65%).

An analysis of sample F10 shows a similar behavior to most of the shales analyzed. There is a considerable SiO_2 content (53.27%), and the presence of Fe_2O_3 (6.10%), which may indicate the presence of illite in the sample, can be corroborated by the presence of K_2O in the percentage of 3.33%. The presence of SO_3 was also observed, with a content of 3.08%, which is a considerable value and can probably be detected by means of the sample's differential thermal analysis curve.

The chemical analysis of samples F10 and F11 was similar, with the latter showing a high percentage of SiO_2 (50.63%), indicating the presence of quartz; the presence of illite can be indicated by the presence of K_2O (3.03%). There is also a considerable SO_3 content (5.69%).

F12 shows similar characteristics to samples F10 and F11. The presence of illite can be indicated by the Fe_2O_3 content of over 6%. This is reinforced by the presence of a significant amount of K_2O. A MgO content of around 2% was also observed, which may indicate the presence of carbonate in the sample. The F13 shale sample also has a high SiO_2 content, indicating the strong presence of quartz in its composition. There are also considerable levels of alkaline oxides (MgO and CaO), which may indicate the presence of carbonate, as well as Fe2O3 and $_{K2O}$, which indicate the presence of illite in the sample.

The Brasgel PA sample has high levels of SiO2 (64.11%) and Al2O3 (18.53%), which indicates the presence of clay minerals such as smectite, while the high Fe2O3 content (9.39%) also indicates the presence of clay minerals from the smectite group.

The Cloisite sample has a very similar chemical composition to the Brasgel PA sample, but it has a higher Al2O3 content, which may indicate a greater quantity of clay minerals such as smectite, illite and kaolinite. It can be seen that the Cloisite clay has a considerably lower Fe2O3 content than the Brasgel PA clay, which may be related, according to Silva *et al.* (2012), to the Cloisite clay processing process.

In a study carried out by Silva *et al.* (2012), in relation to the chemical analysis of samples of Brasgel PA and Closite clays, it was observed that the chemical composition obtained for these samples were consistent with those obtained in the analysis in Table 7, even indicating the same differences between the analyses of the samples, i.e. higher $_{SiO2}$, $_{Fe2O3}$, CaO, MgO and $_{K2O}$ contents for Brasgel PA clay compared to Cloisite clay, and higher $_{Al2O3}$ and $_{Na2O}$ contents for Cloisite clay, confirming the results in Table 7 for samples of the same clay species.

As can be seen, the chemical analysis provides an indication of the presence of some clay minerals, as well as some impurities in the samples studied. The X-ray diffraction results can confirm these indications and will be discussed in section 4.1.5.

4.1.4 Differential thermal analysis and thermogravimetry

Figures 20 to 34 show, respectively, the thermodifferential and thermogravimetric analysis curves for the samples of shales F1 to F13 and the reactive clays Brasgel PA and Cloisite. The results of the quantitative analysis of the thermogravimetry curves for the samples studied can be seen in Table 9.

Figura 20 - Curves of: a) differential thermal analysis and b) gravimetric thermal analysis for the F1 shale sample.

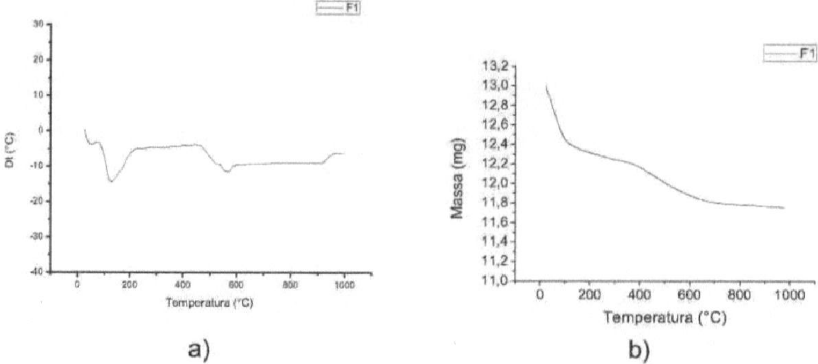

Figura 21 - Curves of: a) differential thermal analysis and b) gravimetric thermal analysis for the F2 shale sample.

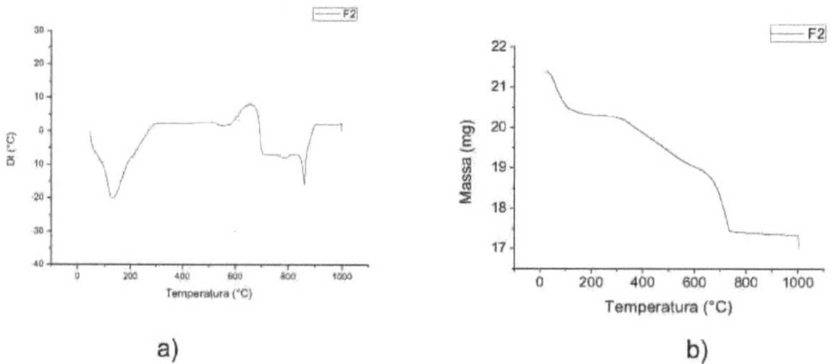

Figura 22 - Curves of: a) differential thermal analysis and b) gravimetric thermal analysis for the F3 shale sample.

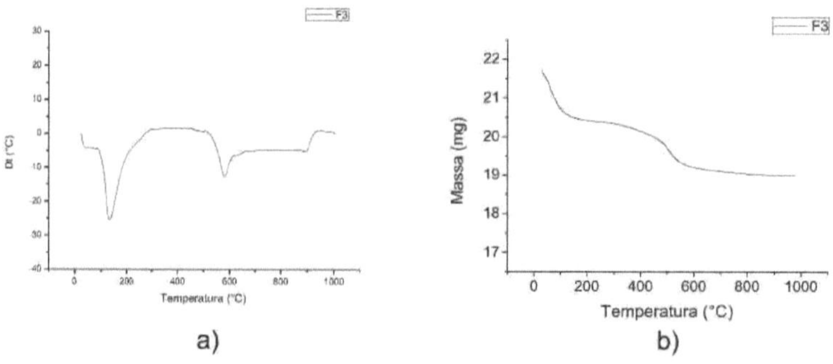

Figura 23 - Curves of: a) differential thermal analysis and b) gravimetric thermal analysis for the F4 shale sample.

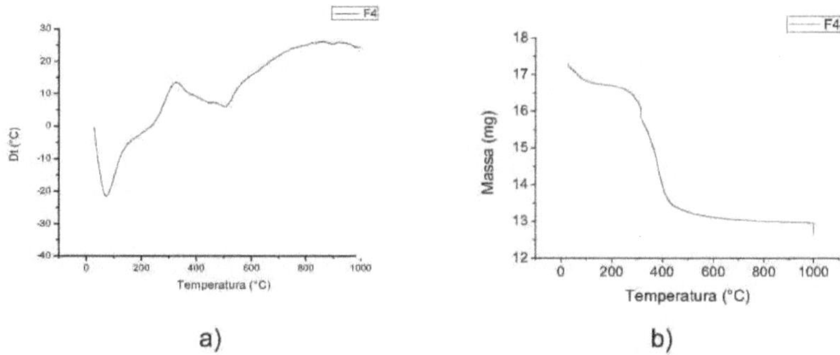

a)   b)

Figura 24 - Curves of: a) differential thermal analysis and b) gravimetric thermal analysis for the F5 shale sample.

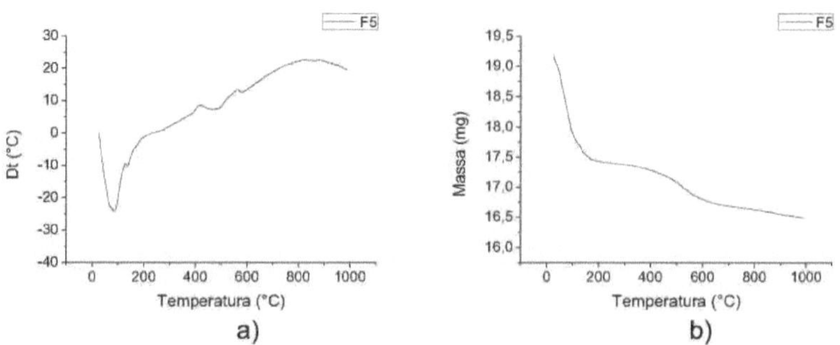

a)   b)

Figura 25 - Curves of: a) differential thermal analysis and b) gravimetric thermal analysis for the F6 shale sample.

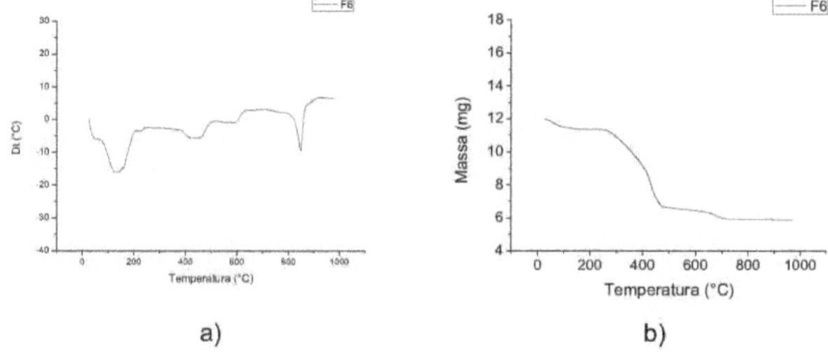

a)   b)

Figura 26 - Curves of: a) differential thermal analysis and b) gravimetric thermal analysis for the F7 shale sample.

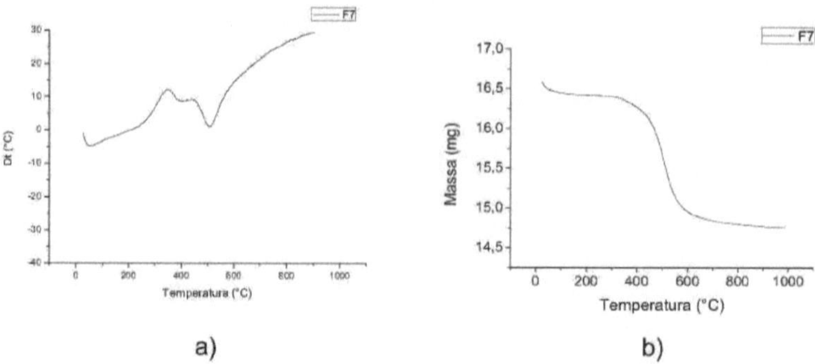

a)

b)

Figura 27 - Curves of: a) differential thermal analysis and b) gravimetric thermal analysis for the F8 shale sample.

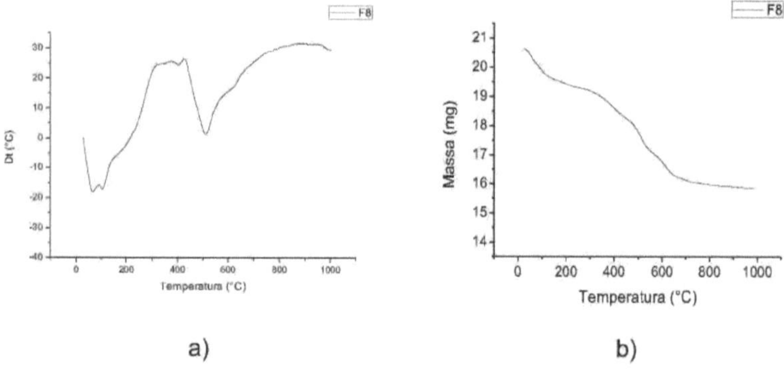

a)

b)

Figura 28 - Curves of: a) differential thermal analysis and b) gravimetric thermal analysis for the F9 shale sample.

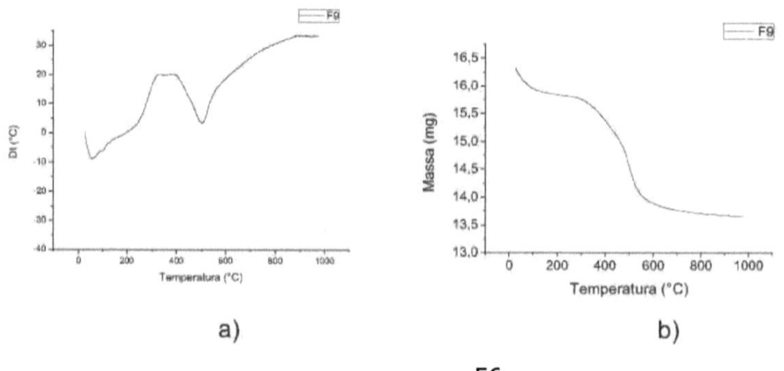

a)

b)

Figura 29 - Curves of: a) differential thermal analysis and b) gravimetric thermal analysis for the F10 shale sample.

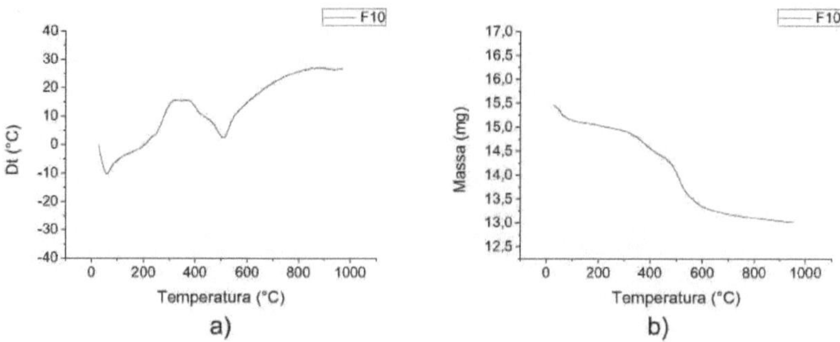

Figura 30 - Curves of: a) differential thermal analysis and b) gravimetric thermal analysis for the F11 shale sample.

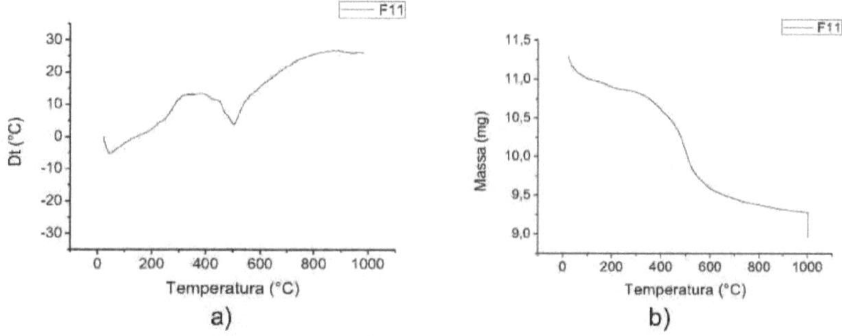

Figura 31 - Curves of: a) differential thermal analysis and b) gravimetric thermal analysis for the F12 shale sample.

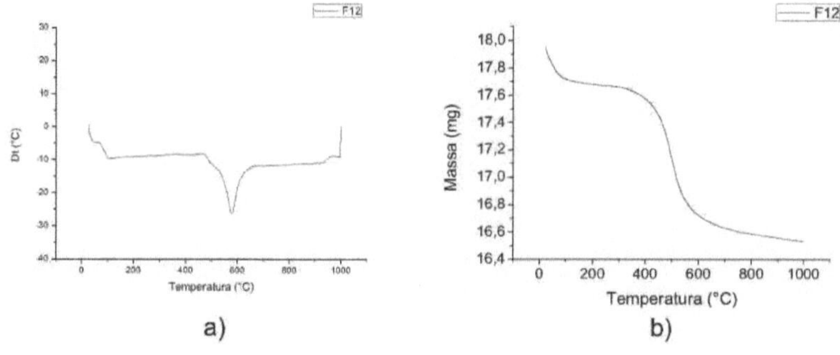

Figura 32 - Curves of: a) differential thermal analysis and b) gravimetric thermal analysis for the F13 shale sample.

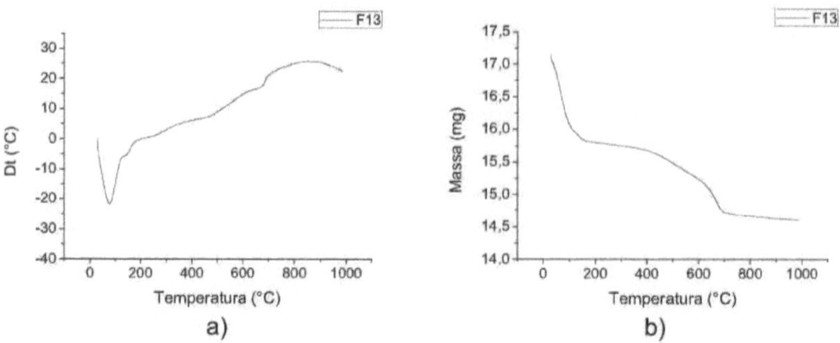

a)   b)

Figura 33 - Curves of: a) differential thermal analysis and b) gravimetric thermal analysis for the Brasgel PA sample.

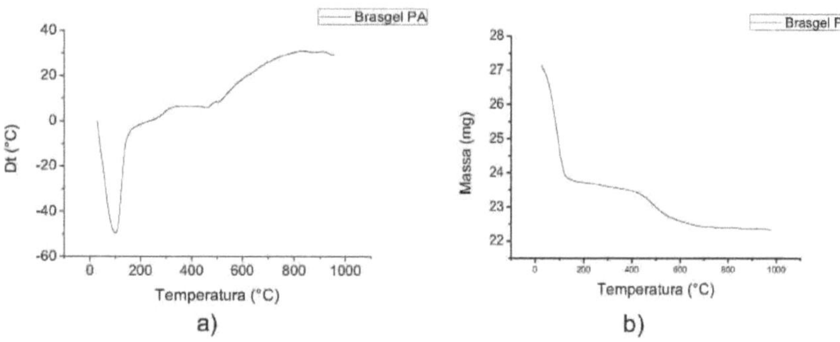

a)   b)

Figura 34 Curves of: a) differential thermal analysis and b) gravimetric thermal analysis for the Cloisite sample.

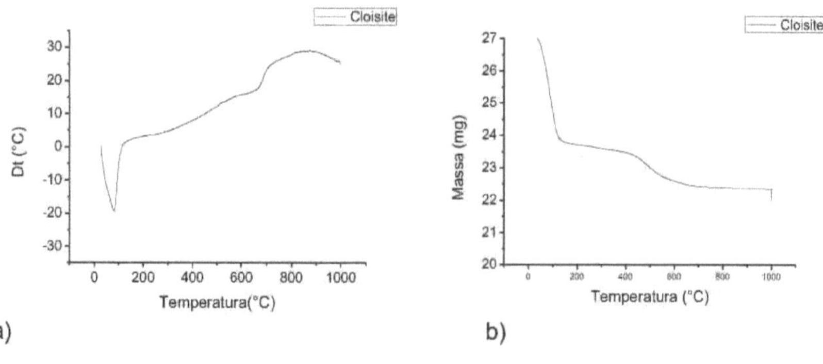

a)   b)

Figure 20a shows an endothermic peak at 125^0 C, which most likely indicates the presence of adsorbed water. There is an exothermic band with a peak at 5200C, which may be related to the presence of hydroxyl.

The graph in Figure 20b shows a total loss of mass of 12.11%. The first loss observed is most likely due to the loss of free and adsorbed water. There is also a second inflection between around 450 and 650° C which is probably related to the loss of hydroxyl.

Figure 21a shows the ATD curve for sample F2. Firstly, there is an endothermic peak at around 130oc which characterizes the presence of adsorbed water. An exothermic peak is observed at around 650° C, referring to hydroxyl. Finally, another endothermic peak is detected at around 860oc which may indicate the presence of magnesium carbonate ($MgCO_3$), characterizing the existence of carbonate in the sample.

Figure 21b shows a total mass loss of 20.59%. The first loss is most likely due to the loss of free and coordinated water and the second inflection is due to the loss of organic matter and hydroxyls.

The ATD curve for sample F3 in Figure 22a shows an endothermic peak at around 132oc related to water loss. The endothermic inflection at around 58oc may indicate the presence of hydroxyl. Figure 22b shows a total mass loss of 15.98 %. The first loss represents free and coordinated water and the second loss found is most likely due to the loss of hydroxyls.

In relation to the ATD curve in Figure 23a, there is first an endothermic peak at around 90oc related to the loss of free water. Around 310oc there is an exothermic peak characteristic of the combustion of organic matter. The endothermic inflection at around 500oc may indicate the presence of hydroxyl. A total mass loss of 24.72% is observed in the thermogravimetry test shown in Figure 23b. From this last figure, we can see a small inflection relating to the loss of water, a second inflection which represents a more evident loss, which is most likely related to mass losses due to the combustion of organic matter, the total mass loss of the sample was 24.72%.

In relation to Figure 24a, there is an endothermic peak at 100° C related to adsorbed water and a "shoulder" at approximately 200oc which may be indicative of the presence of water coordinated to the magnesium cations. At around 400oc, it refers to the combustion of organic matter and the next band refers to the presence of hydroxyl. Figure 24b shows a total mass loss of 15.66%. This loss was close to what is observed in reactive clays. The first loss, which represents the highest percentage of mass lost, probably refers to the loss of free and coordinated water, while the second loss is probably related to organic matter and hydroxyls.

In Figure 25a, there is an endothermic inflection at around 130oc referring to the slight presence of water. The second peak observed at around 450oc probably refers to the combustion of organic matter. An exothermic peak is observed at around 678° C, referring to hydroxyl. Finally, another endothermic peak is detected at around 848oc which may indicate the presence of magnesium carbonate ($MgCO_3$), characterizing the existence of carbonate in the sample. Figure 25b shows a very significant total loss of mass of around 53%. This loss is largely due to the combustion of organic matter and sulphides; first there is a loss of mass related to the loss of water, then there is a loss of mass due to combustion and the loss of hydroxyl, as mentioned above.

Figure 26a shows the differential thermal analysis curve for sample F7. There is an endothermic inflection due to the presence of adsorbed water, followed by an exothermic peak at around 350oc, which probably refers to the combustion of organic matter, which generally occurs between 200 and 400oc. The peak observed at 500oc refers to the loss of hydroxyl. Figure 26b shows a total mass loss of 12.9%. There is a less pronounced loss of water, followed by a loss of mass associated with the combustion of organic matter and the loss of hydroxyl.

The ATD curve in Figure 27a shows an endothermic peak related to the presence of adsorbed water, an exothermic band between around 280° C and 430OC, which may be related to the combustion of organic matter and a new exothermic inflection probably related to the combustion of sulphide, since the sample, according to the XRF results, has a high SO_3 content (9.87%). At around 520OC an endothermic peak is observed, which may be related to the presence of hydroxyl. In relation to Figure 27b, the first loss of mass was due to adsorbed and coordinated water and the second loss was due to the combustion of organic matter and the loss of hydroxyl. The total mass loss was 24.72%.

Figure 28a shows an endothermic peak related to the presence of adsorbed water, followed by an exothermic band, which is probably related to the combustion of organic matter followed by the combustion of sulphides. A peak is observed at approximately 510OC, which may indicate the presence of magnesium hydroxide or even the loss of hydroxyl. An analysis of Figure 28b shows a total mass loss of 18.37%. The first mass loss refers to free water and the second loss comes from the combustion of organic matter and SO_3.

From the curve shown in Figure 29a, there is an inflection at around 80OC referring to the presence of water and an exothermic band between around 290OC and 400OC which may be related to the combustion of organic matter and sulphide. At around 520OC an endothermic peak is observed, which may be related to the presence of hydroxyl. Figure 29b shows behavior similar to that observed for sample F9 in Figure 30b, with a total loss of 17.94%.

For the curve shown in Figure 30a, there is an endothermic inflection related to the presence of water, an exothermic band is found at around 380OC which may be related to the presence of organic matter and a peak at 500OC referring to the presence of hydroxyl. The curve shown in Figure 30b shows two points of mass loss: the first refers to the loss of water and the second refers to the combustion of organic matter and the loss of hydroxyl; these losses total 16.76% of the original mass of the sample.

For the curve in Figure 31a, an endothermic inflection is observed in relation to the presence of water. An endothermic inflection is observed at around 579^0 C and this must refer to the presence of hydroxyl. The total mass loss observed in Figure 31b was 7.92% of the total mass.

The ATD curve in Figure 32a shows an endothermic peak due to the presence of adsorbed water, an endothermic band characteristic of the presence of structural hydroxyls and iron-rich clays and an endo-exothermic peak characteristic of the destruction of the crystalline lattice. Figure 32b shows a total mass loss of 17.76%, the first loss of mass being due to adsorbed and coordinated water and the second to hydroxyl loss.

For the curve shown in Figure 33a, there is an endothermic inflection related to the presence of water at around 10Q0C, followed by an endothermic inflection at around 7000C, which is probably linked to the presence of hydroxyl. For the curve shown in Figure 33b, there are two points of mass loss: the first refers to the loss of water and the second, less obvious, refers to the loss of hydroxyl; these losses total 16.67% of the original mass of the sample.

There is a great deal of similarity between the ATD curves of the samples shown in Figures 33a and 34a. This is due to the proximity of the chemical composition and characteristics of both samples. The loss of mass observed for the Brasgel PA and Cloisite samples in Figures 33b and 34b also point to a very similar value, suggesting that they both behave very similarly.

Table 9 - Quantitative analysis of the thermogravimetry curves for the samples studied.

SAMPLE NAME	Mass loss (mg)	Total mass loss (%)
F1	1,58	12,11
F2	4,41	20,59
F3	3,33	15,38

F4	4,64	26,23
F5	3,00	15,66
F6	6,47	53,81
F7	2,14	12,9
F8	5,10	24,72
F9	3,00	18,37
F10	2,66	17,94
F11	2,86	16,76
F12	1,42	7,92
F13	2,33	20,65
Brasgel PA	1,38	17,76
Cloisite	4,50	16,67

4.1.5 X-ray diffraction

The diffractograms of the samples studied can be seen in Figures 35 to 49. The X-ray diffraction tests were carried out on the samples without and with the presence of ethylene glycol. For the latter, the samples have the acronym EG after their normal nomenclature.

The symbols on the graphs refer to the following minerals: E - Smectite Group; I - Illite; C - Kaolinite; Q - Quartz; D - Dolomite.

Figure 35- X-ray diffractogram for the F1 shale sample with and without ethylene glycol.

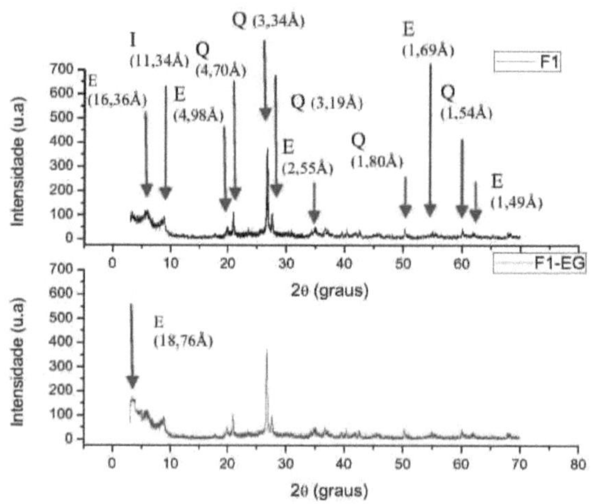

Figure 36- X-ray diffractogram for the F2 shale sample with and without ethylene glycol.

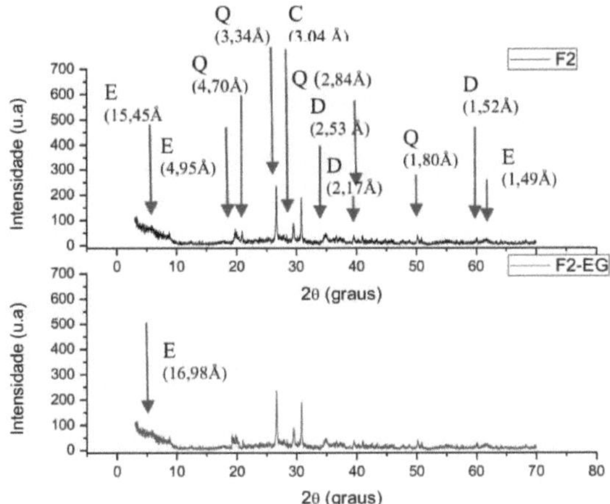

Figure 37- X-ray diffractogram for the F3 shale sample with and without ethylene glycol.

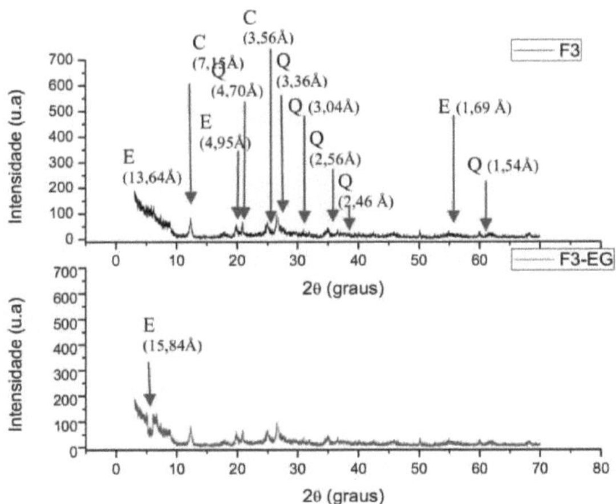

Figure 38- X-ray diffractogram for the F4 shale sample with and without ethylene glycol.

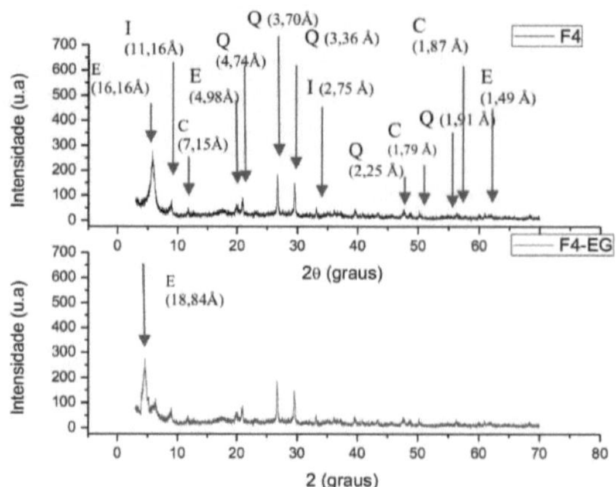

Figure 39- X-ray diffractogram for the F5 shale sample with and without ethylene glycol.

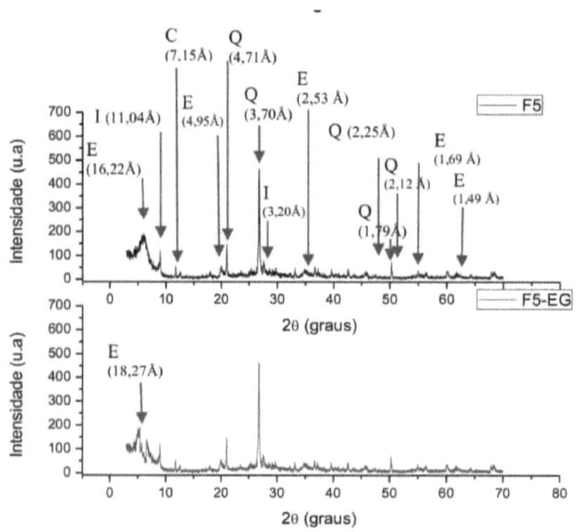

The diffractogram in Figure 35 of the F1 shale shows a peak at 16.36A with a change to 18.76A after treatment with ethylene glycol. There are also peaks at 4.98A, 2.55A, 1.69A and 1.49A referring to smectite, a peak at 11.34A probably referring to the presence of illite, peaks at 4.70A, 3.70A, 3.34A, 3.19A, 1.80A and 1.54A are also highlighted and these must be related to the presence of quartz in the sample analyzed.

Figure 36 shows peaks at 15.45A, 4.95A and 1.49A, which are characteristic of the presence of clay minerals from the smectite group in the sample. The high levels of magnesium and calcium oxides come from dolomite, as indicated by the chemical composition analysis.

Peaks referring to dolomite can be observed at 2.53A, 2.17A and 1.52A. The presence of quartz was evidenced by the peaks at 4.70A, 3.34A, 3.04A, 2.84A and 1.80A. The characteristic peak of smectite for the sample treated with ethylene glycol was shifted to 16.98 A .

In the diffractogram of the F3 shale shown in Figure 37, the peak shifted from 13.64A to 15.84A for the sample treated with ethylene glycol. This shift is characteristic of the presence of clay minerals from the smectite group. Peaks at 4.95A and 1.69A also characterize the presence of argillominerals from the smectite group. There are also peaks at 4.70A, 3.36A, 3.04A, 2.56A and 1.54A characteristic of the presence of quartz, as well as peaks at 7.15A and 3.56A related to the presence of kaolinite in the sample.

The diffractogram of shale F4, shown in Figure 38, shows a peak at 16.16A with a shift of 18.84A, which characterizes the presence of a clay mineral from the smectite group, peaks at 11.16A and 2.75A, which are characteristic of the presence of illite, peaks at 7.15A, 1.87A and 1.79A indicating the presence of kaolinite, peaks at 4.98A and 1.49A confirming the presence of the smectite group and peaks at 3.70A, 3.36A, 2.25A and 1.91A referring to the presence of quartz.

Figure 39 shows the diffractogram of the F5 shale. You can see peaks that characterize the presence of smectite in its composition, as can be seen in the shift of the peak from 16.22A to 18.27A for the sample treated with ethylene glycol. The peaks at 4.95A, 2.53A, 1.69A and 1.49A are also related to the presence of clay minerals from the smectite group. Peaks at 11.04A and 3.20A refer to the presence of illite. The presence of quartz can be confirmed by the presence of peaks at 4.70A, 3.70A, 2.25A and 1.79A.

Figure 40- X-ray diffractogram for the F6 shale sample with and without ethylene glycol.

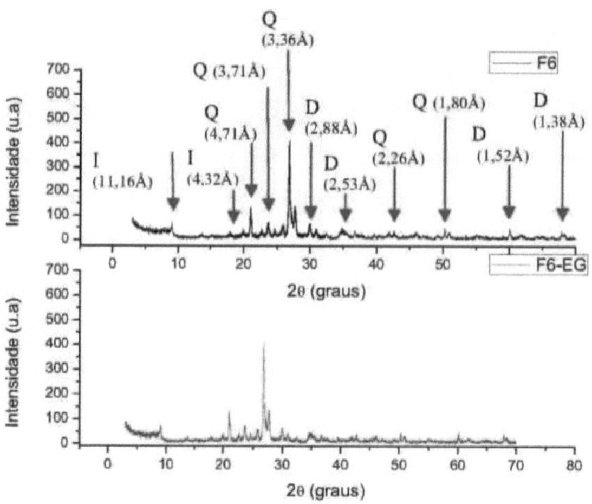

Figure 41- X-ray diffractogram for the F7 shale sample with and without ethylene glycol.

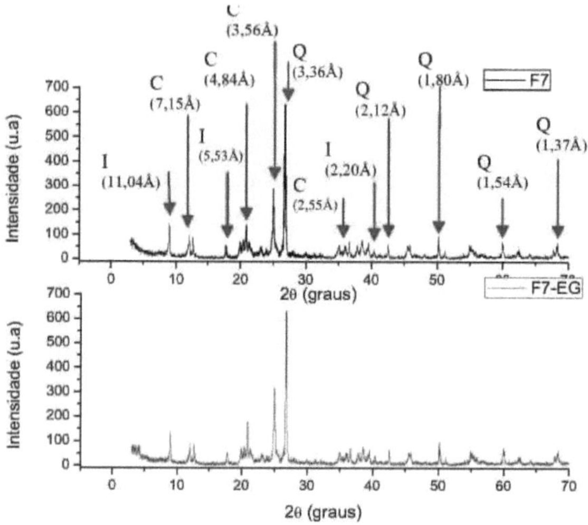

Figure 42- X-ray diffractogram for the F8 shale sample with and without ethylene glycol.

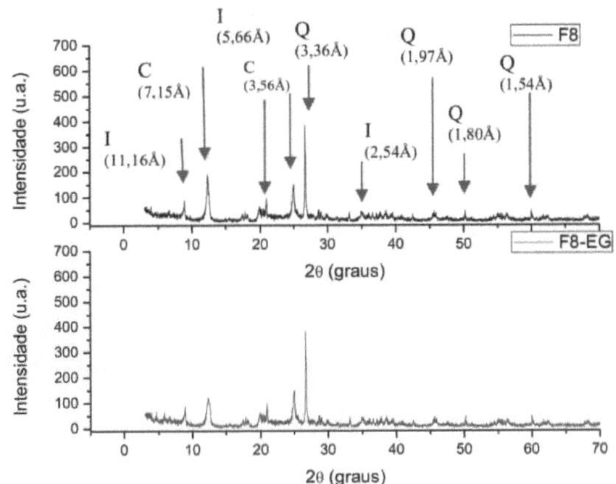

Figure 43- X-ray diffractogram for the F9 shale sample with and without ethylene glycol.

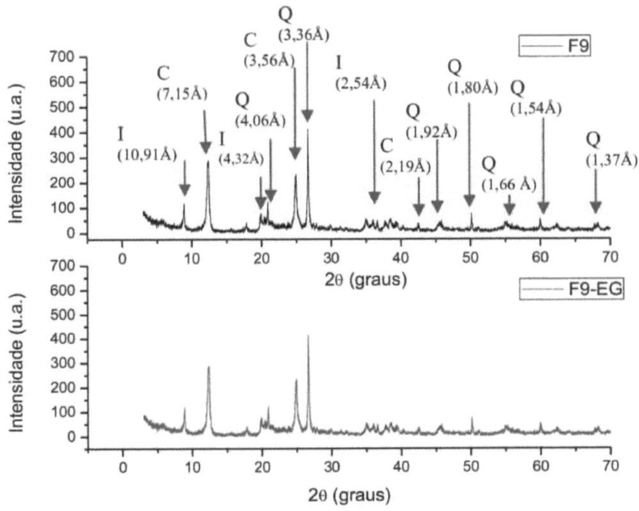

Figure 40 shows peaks that characterize a very varied composition. This diffractogram shows peaks at 11.04A and 4.32A that are typically related to the presence of illite. The high calcium content detected in the chemical analysis may be related to the presence of dolomite observed by the peaks at 2.88A, 2.53A, 1.52A and 1.38A, as well as peaks at 4.71A, 3.71A, 3.36A and 1.80A, which are characteristic of the presence of quartz.

The diffractogram in Figure 41 shows the absence of peaks that characterize the presence of smectite in the composition of the F7 formation, confirmed by the absence of a peak shift in the presence of ethylene glycol in the sample, for this sample, peaks are observed at

11.04A, 5.53A and 2.20A, characteristic of the presence of illite, at 7.93A, 4.84A, 3.56A and 2.55A indicating the presence of kaolinite, and at 3.36A, 2.12A, 1.80A, 1.54A and 1.37A characteristic of the presence of quartz.

In the diffractogram shown in Figure 42, there is no peak shift for the sample with ethylene glycol, which indicates the absence of clay minerals from the smectite group in the sample, peaks at 11,16A, 5.66A and 2.54A, referring to illite, peaks at 7.37A and 3.56A indicating the existence of kaolinite, and the peaks at 3.69A, 1.97A, 1.80A and 1.54A confirm the presence of quartz. The diffractogram in Figure 43 shows the presence of illite in the sample, characterized by the peaks at 10.91A and 2.54A, the presence of kaolinite is detected by the peaks at 7.39A and 3.53A, and the presence of quartz is also evidenced by the peaks at 4.06A, 1.92A, 1.80A and 1.54A.

Figura 44- X-ray diffractogram for the F10 shale sample with and without ethylene glycol.

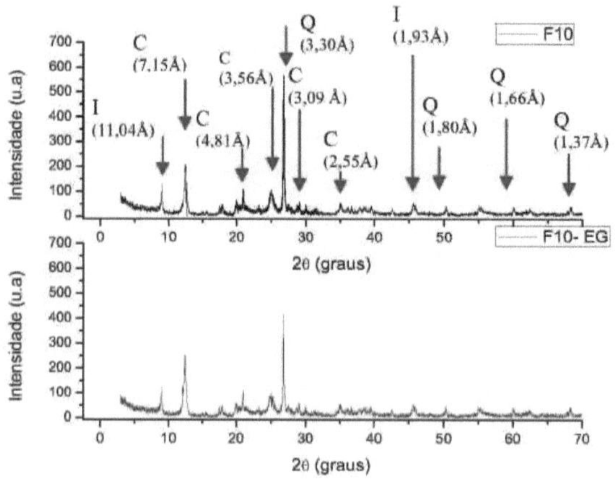

Figura 45-   X-ray diffractogram for the F11 shale sample with and without ethylene glycol.

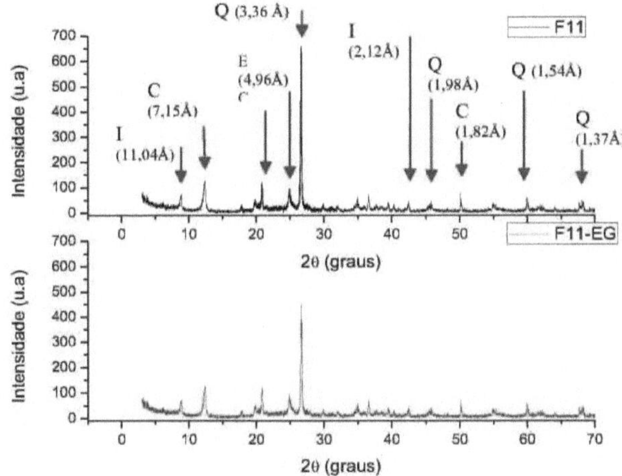

Figura 46- X-ray diffractogram for the F12 shale sample with and without ethylene glycol.

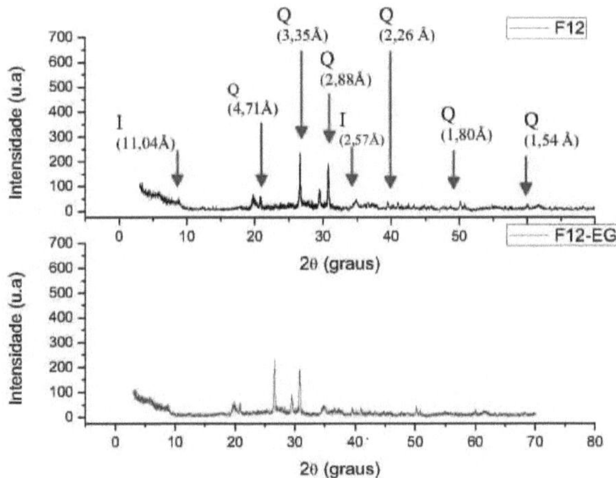

Figura 47-   X-ray diffractogram for the F13 shale sample with and without ethylene glycol.

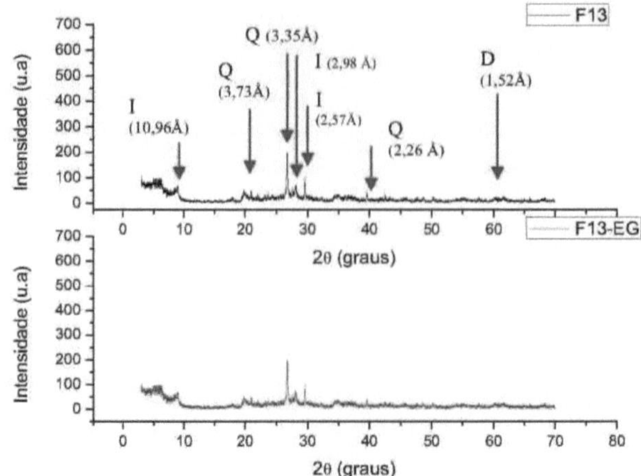

Figura 48- X-ray diffractogram for the Brasgel PA clay sample with and without ethylene glycol.

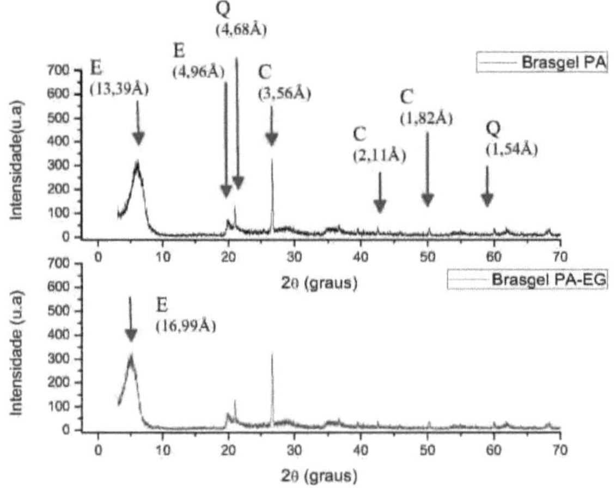

Figure 49 - X-ray diffractogram for the Cloisite clay sample with and without ethylene glycol.

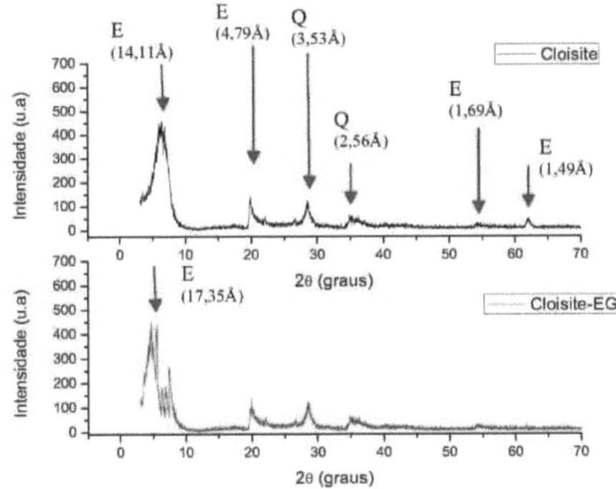

Analysis of the diffractogram in Figure 44 for sample F10 shows the absence of a peak that characterizes the presence of clay minerals from the smectite group, so there is no change in peaks in the diffractogram of the sample with ethylene glycol. The 11.04A and 1.93A peaks characterize the presence of illite in the sample, peaks at 7.93A, 4.81A, 3.56A, 3.09A and 2.55A for kaolinite, and the presence of peaks at 3.30A, 1.80A, 1.66A and 1.37A characterize the presence of quartz.

The analysis of Figure 45 of sample F11 shows the presence of illite, with peaks at 11.04A and 2.12A, quartz through the detection of peaks at 3.36A, 1.98A and 1.54A and kaolinite with peaks at 7.39A, 3.56A and 1.82A. Samples F12 and F13 in the diffractograms in Figures 46 and 47 follow the same behavior observed for samples F11 and F12.

From the diffractogram in Figure 48 referring to the analysis of the Brasgel PA clay sample, we can see that the peak has shifted from 13.39A to 16.99A, characterizing the presence of clay minerals from the smectite group. This is also evidenced by the peak at 4.96A, also characteristic of the smectite group, and the peaks at 4.68A, 1.80A and 1.54A referring to the presence of quartz. Studies carried out by Menezes *et. al.* (2009) and Batista *et. al.* (2009) identified mineralogical phases (smectite, kaolinite and quartz) compatible with those observed in the diffractogram of the Brasgel PA clay sample shown in Figure 48, thus indicating that the Brasgel PA sample showed a diffractogram typical of a bentonite clay.

Analysis of the diffractogram of the reactive clay Cloisite in Figure 49 clearly shows peaks characteristic of smectite (shift from 14.41A to 17.35A and peak at 4.79A). The presence of kaolinite is evidenced by the peaks at 2.11A and 1.82A and there are also peaks at 3.53A, 2.56A and 1.54A relating to quartz.

71

The XRD of the Cloisite sample shown in Figure 49 indicates, as does the XRD of the Brasgel PA sample, the presence of smectite-type clay and quartz. This shows that both are reactive clays with a high degree of expansion, and this hydration potential is probably due to the presence of smectite in their composition. Studies by Sousa Santos (1992) and Menezes *et al.* (2009) show that the diffractograms in Figures 48 and 49 are typical of bentonite clays.

The XRD tests indicated that the F1, F2, F3, F4 and F5 shales show peaks characteristic of the presence of clay minerals from the smectite group, as do the reactive clays Brasgel PA and Cloisite. The displacement of the peak obtained for the tests carried out with ethylene glycol was observed in samples F1, F3, F4, F5 and samples of Brasgel clay and Cloisite.

A joint analysis of the results obtained from the shale characterization tests shows that the shales F1, F2, F3, F4 and F5, as well as the reactive clays Brasgel PA and Cloisite, are indicative of reactivity, since they have smectitic clay minerals in their composition. Therefore, these samples are likely to be the most reactive when it comes to hydration.

It can also be concluded that samples F6, F7, F8, F9, F10, F11, F12 and F13 are indicative of low reactivity to expansion, since they were found to have low CTC and AE values, as well as XRF, ATD, TG and XRD results which do not indicate the presence of clay minerals from the smectite group. Therefore, as the reactivity of shales is basically linked to their content of reactive clay minerals, it can be pointed out that these samples do not have a high degree of hydration according to the characterization tests carried out.

4.2 Selecting the best concentrations of clay swelling inhibitors

4.2.1 Clay swelling

Figures 50 and 51 show the results of the swelling tests of the inhibitors potassium sulphate, potassium acetate, potassium citrate and potassium chloride (KCl) alone, for three different concentrations (16, 18 and 20g/350ml of water) in the presence of Brasgel PA and Cloisite clays, respectively.

The classifications for swelling were: values equal to or less than 2mL/g were considered to be non-swelling or zero swelling, values greater than 2 and less than or equal to 5mL/g were considered to be low swelling, values greater than 5 and less than or equal to 8mL/g were considered to be medium swelling and values above 8mL/g were considered to be high swelling (FERREIRA, 2009).

Figura 50- Swelling applied to Brasgel PA clay for the inhibitors potassium sulphate, potassium acetate, potassium citrate and potassium chloride at concentrations of 16, 18 and

20g of inhibitor/350mL of water.

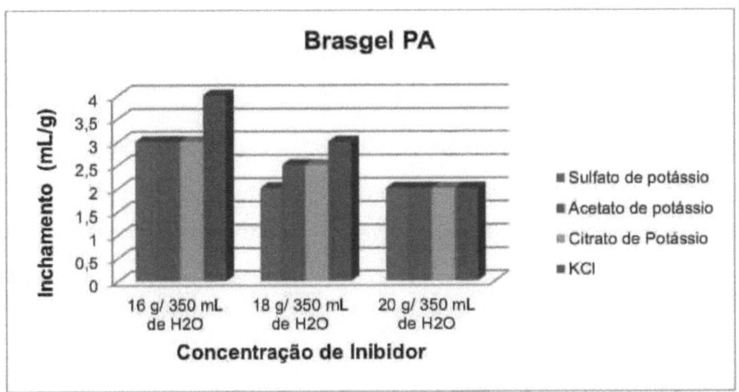

Figura 51- Swelling applied to Cloisite clay for the inhibitors potassium sulphate, potassium acetate, potassium citrate and potassium chloride at concentrations of 16, 18 and 20g of inhibitor/350mL of water.

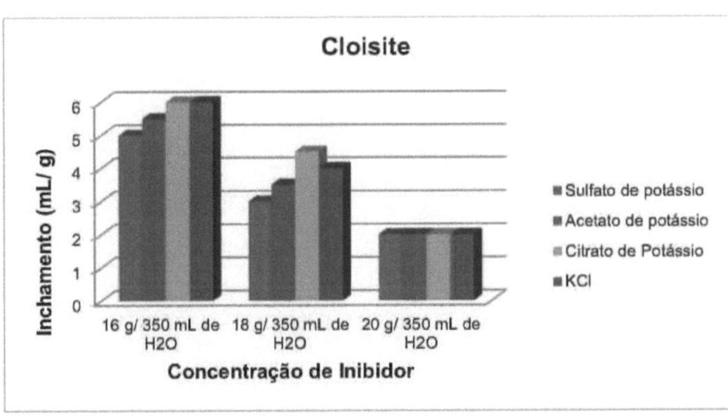

Looking at Figure 50, it can be seen that the concentration of 18g of potassium sulphate/350mL of water already shows zero swelling, proving effective inhibition from this concentration onwards. It can also be seen that at a concentration of 20g/350mL all the inhibitors analyzed showed zero swelling, in accordance with the classification established by the test. For the test in water (without the presence of an inhibitor) Brasgel PA showed a swelling of 16.5mL/g of clay. This behavior is associated with the high chemical activity of the water, which generates an osmotic flow that promotes the migration of water from the aqueous solution to the clay under study.

The graph in Figure 51 shows that only at a concentration of 20g/350mL of water is there no expansion for Cloisite clay when compared to all the inhibitors. This can be explained by the high expansion capacity of Cloisite clay, which makes it more difficult to control hydration, thus requiring a higher concentration of inhibitors to effectively control hydration of this sample. The swelling obtained for the Cloisite test in water was 24mL/g of clay.

The need for an increase in inhibitor concentration for effective expansion control may be related to what was already pointed out in the literature review, i.e. chemical activity, as seen in Figure 6. Therefore, for the drilling fluid to have a lower activity than the formation, it is necessary for it to have a certain concentration of salt, so that the flow of water occurs from the formation to the drilling fluid, thus avoiding expansion.

The phenomenon of chemical diffusion is related, according to Yan and Deng (2013), to the difference in chemical potential between the salt solution and the formation subjected to this solution. There is thus a migration of solute from the zones of high concentration (salt solution) to the formation, i.e. chemical diffusion may dominate the migration of solute to the shales studied. From this, it can be indicated that the inhibition observed by the swelling test is probably associated with this mechanism, i.e. the ions that diffuse from the solution into the formation interact favorably with these formations, thus reducing their swelling, and the intercalation of these diffused ions hinders the penetration of water, preventing their swelling.

From the above and the analysis of the test results, it can be seen that the concentration of 20g of inhibitor/350mL of water as a component of the solution guarantees a sufficient difference in concentration to prevent the invasion of water from the fluid into the samples, which explains the nullity of swelling for this concentration for all the inhibitors studied.

In the joint analysis of the results obtained for hydration control of Brasgel PA and Cloisite clays, it can be seen that in both cases the potassium citrate inhibitor showed a significant increase in its hydration control capacity as the concentration of the inhibitor increased. In this way, this inhibitor showed a very strong ability to reduce expansion with a progressive increase in concentration.

The high hydration control capacity of this inhibitor may be due to the physical inhibition promoted by the counter-ion, which has three anionic active sites as can be seen in Figure 10, presented in the literature review, which promotes an increase in physical inhibition (from the joining of the clay edges) with increasing concentration. In addition, all the inhibitors were effective in controlling swelling when compared to the tests carried out without the presence of these additives.

Based on the results obtained, it can be indicated that for this test the concentration that showed the most positive results for the preparation of drilling fluids was 20g of inhibitor/350mL of water, since for this concentration all the inhibitors studied showed zero swelling.

4.2.2 Determination of free water in clay samples by capillary suction

Figures 52 and 53 show the results of the suction time determination tests for solutions containing the inhibitors potassium sulphate, potassium acetate, potassium citrate and potassium chloride (KCI) alone, for three different concentrations (16, 18 and 20g/350ml of water) in relation to Brasgel PA and Cloisite clays, respectively. From these figures, it is possible to indicate which additive has the best inhibition capacity in relation to this parameter, because the shorter the suction time, the greater the amount of free water and, consequently, the lower the water-clay interaction. It was also determined that the suction time values for solutions without the presence of an inhibitor were 121 seconds for Brasgel PA clay and 154 seconds for Cloisite clay.

The high capillary suction time shown for the samples submitted to the test indicates that the absence of the inhibiting salt makes the interaction between the water and the clay very strong. According to Ewy and Stankovich (2002), this expansion phenomenon occurs due to the absence of the potassium ion, which allows water to easily enter the clay layers, causing them to delaminate and their layers to separate.

Figura 52-   Capillary suction time in relation to Brasgel PA clay for the inhibitors potassium sulphate, potassium acetate, potassium citrate and potassium chloride at concentrations of 16, 18 and 20g of inhibitor/350mL of water.

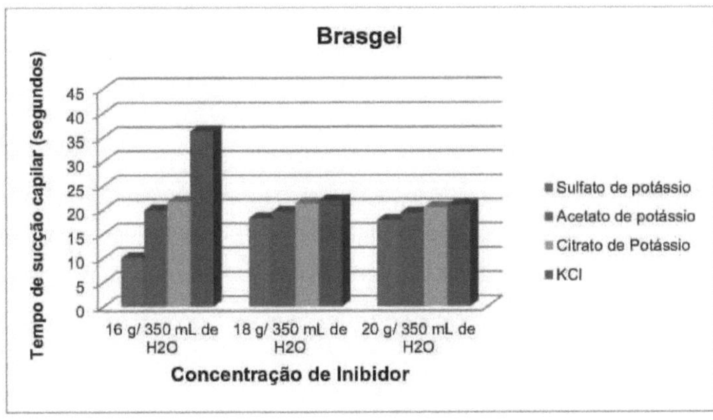

Figura 53- Capillary suction time in relation to Cloisite clay for the inhibitors potassium sulphate, potassium acetate, potassium citrate and potassium chloride at concentrations of 16, 18 and 20g of inhibitor/350mL of water.

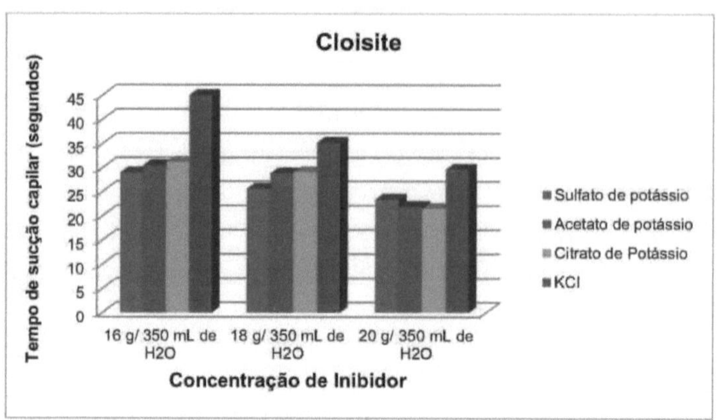

The importance of osmotic pressure (the pressure that prevents the osmosis process) is highlighted to help understand the results obtained in the capillary suction free determination test. Based on the calculations carried out, it can be seen that potassium sulphate is, among all the inhibitors, the one with the lowest osmotic pressure, so this inhibitor makes it possible for osmosis to occur, resulting in the passage of water inside the samples analyzed into the drilling fluid, thus obtaining less swelling. This is probably the reason why this inhibitor has the shortest capillary suction times.

Potassium chloride, on the other hand, has much higher osmotic pressures compared to those obtained for the other inhibitors studied; this fact explains its inferior performance in controlling the hydration of the samples, thus presenting the longest capillary suction times.

Figure 52 shows that the suction time decreased as the concentration of the inhibitors increased. This behavior was to be expected given that increasing the concentration of the inhibitor in the solution makes a greater quantity of K ions available[+] . According to Ye *et al.* (2010), this increase hinders the delamination of the clay particles by controlling the hydration that this ion promotes. The three chlorine-free inhibitors (potassium sulphate, potassium acetate and potassium citrate) showed more positive results in controlling the expansion of Brasgel PA clay than those observed for the potassium chloride inhibitor (a product widely used by the oil industry). One explanation for this result could be the fact that chlorine salts act to increase the density of the solution, resulting in an increase in suction time.

It is known that the transfer of potassium ions to the layers of reactive formations can reduce the distance between the clay plates, leading to swelling retraction. In this way, it is clear that higher salt concentrations promote lower water activity, resulting in a shorter capillary suction time, since there is a greater amount of free water due to the chemical differential hindering the influx of water into the layers of reactive formations.

It was also observed that the solutions containing potassium citrate as a swelling inhibitor promoted a clear progressive reduction in swelling compared to the other inhibitors studied. The same behavior was observed in relation to the swelling tests, as discussed above. Solutions containing potassium chloride as an inhibitor had the longest suction times (36.2; 21.9 and 20.9 seconds for the concentrations of 16, 18 and 20g of inhibitor/350mL of water, respectively), i.e. the smaller the amount of free water present, caused by a greater interaction between the water and the clay and, for this reason, these inhibitors do not act efficiently when compared to the other compositions containing the chlorine-free inhibitors. It should also be noted that all the inhibitors, at all the concentrations studied, showed much lower suction time results when compared to exposing the clay to a solution composed only of water, which confirms the efficient action of the additives in controlling expansion.

The high value of capillary suction time for the solution containing KCl is compatible with the value obtained by Nascimento *et al.* (2009), in which the aqueous solution containing this inhibitor in the presence of reactive clay showed the least effective performance in controlling the interaction of clay with water, i.e. this inhibitor showed the highest capillary suction time among the four inhibitors studied by the authors, with a time of 33.5 seconds for the solution with the lowest concentration and inhibitor.

In general, the same trend was observed for the inhibitor solutions exposed to Cloisite clay, i.e. increasing the inhibitor concentration led to a reduction in suction time, but the suction time values for Cloisite were higher than those obtained for Brasgel PA. This is probably due to the greater reactivity of Cloisite clay, which promotes greater water-clay interaction. Potassium citrate again proved to be the inhibitor that caused the greatest decline in suction times as the concentration of the solution increased.

There was a significant reduction in the suction time values of the solutions containing inhibitors compared to the Cloisite clay solution in the presence of water, which was 154 seconds.

A joint analysis of the results obtained from the capillary suction free water determination tests showed that the inhibitors studied were effective in controlling the hydration of reactive clays. It should also be noted that the chlorine-free inhibitors proved to be more effective in controlling this phenomenon, since they showed lower values for suction time when compared to solutions containing chloride salt. This may also be related to the lower osmotic pressure values of the chloride-free salts, which prevent water from migrating favorably away from the reactive formation.

4.2.3 Benthic inhibition test

The bentonite inhibition test was carried out on Brasgel PA and Cloisite clays and the results can be seen in Figures 54 and 55, respectively. Tables 9 and 10 also show the same test using the viscometer readings at 3 rpm for the bentonite inhibition test for Brasgel PA and Cloisite clay, respectively.

The Tables help to elucidate the results, presenting the reading values on the viscometer for each increase in mass of the clays studied, clearly showing the behavior of each of the inhibitors over the duration of the test. This test simulates the incorporation of clays into a drilling fluid as occurs during the drilling of water-sensitive formations using aqueous drilling fluids.

The performance of the chlorine-free expansion control inhibitors was compared mainly with the performance of potassium chloride. From Figure 54 and Table 10, it can be seen that the chlorine-free inhibitors showed excellent inhibitive properties when compared to potassium chloride. This indicates that the inhibitors potassium sulphate, potassium acetate and potassium citrate have a greater capacity to inhibit reactive formations and this is reflected in the ability of the solutions containing these inhibitors to withstand higher concentrations of clays up to the limiting point of the test (300 rpm).

Figura 54-   Benthic inhibition test in relation to Brasgel PA clay for the inhibitors potassium sulphate, potassium acetate, potassium citrate and potassium chloride.

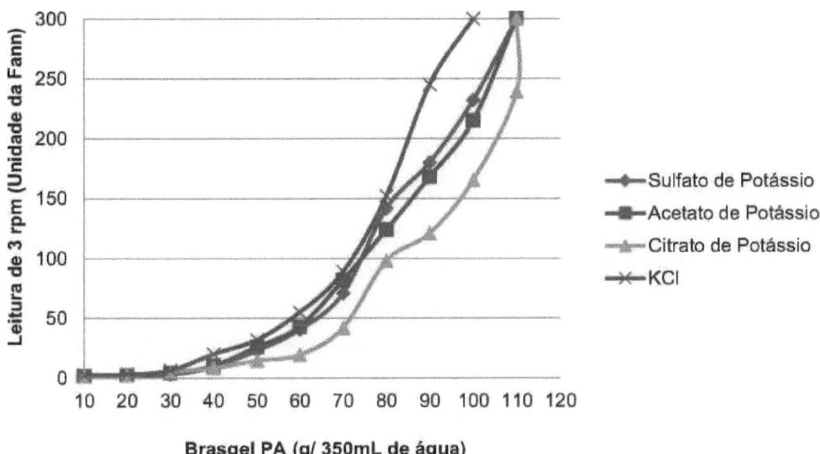

Table 10 - Viscometer readings at 3 rpm for the bentonite inhibition test in relation to Brasgel PA clay for the inhibitors potassium sulphate, potassium acetate, potassium citrate and potassium chloride.

Concentration	Potassium sulphate	Potassium acetate	Potassium citrate	Potassium chloride
10	2,0	1,5	2,0	2,5
20	2,5	2,5	2,5	3,0
30	3,0	4,5	5,5	6,5
40	9,5	10,5	8,5	20,0
50	23,0	26,0	14,5	32,0
60	40,5	43,0	19,5	55,0
70	71,0	82,0	42,0	72,0
80	141,5	124,0	98,0	152,0
90	180,0	168,0	121,0	245,0
100	232,0	215,0	165,0	300,0
110	300,0	300,0	239,0	300,0
120	300,0	300,0	300,0	300,0

It can be seen that an increase in the clay content leads to an increase in the values of the readings obtained for the solution with inhibitor and clay. This behavior is related to the fact that the aqueous solution with salt provides a certain amount of cations that promote hydration inhibition through the adsorption of these cations on the clay surface, However, as the clay content is increased, fewer cations are available in the aqueous solution, resulting in a greater water-clay interaction due to the depletion of cations in the aqueous solution. This result is consistent with the studies presented by Zhong *et al.* (2011), in which the yield of the inhibitors reduced drastically with the progressive increase in clay concentration. Also according to the same author, the use of inhibitors is very effective in the test, since, for the bentonite clay test with water only, the maximum measured value was reached at a clay concentration of 20g/ 350 ml of water.

The results shown in Figure 54 and Table 10 again indicate that potassium citrate was the most effective inhibitor in controlling the hydration of Brasgel PA clay, as it showed lower readings than the other inhibitors for the same gradual increase in the concentration of clay in the solution. This can be explained by the fact that the lower the clay-water interaction, the more free water the solution has. Therefore, the lower the rotation value obtained and, consequently, the greater the effectiveness of the inhibitor used to control expansion.

The better results presented by potassium citrate may be related to the fact that its anionic group reacts differently with clay minerals due to a favorable geometric fit of its structure to the crystalline cross-link units of the clay minerals.

Figura 55-   Benthic inhibition test in relation to Cloisite clay for the inhibitors potassium sulphate, potassium acetate, potassium citrate and potassium chloride.

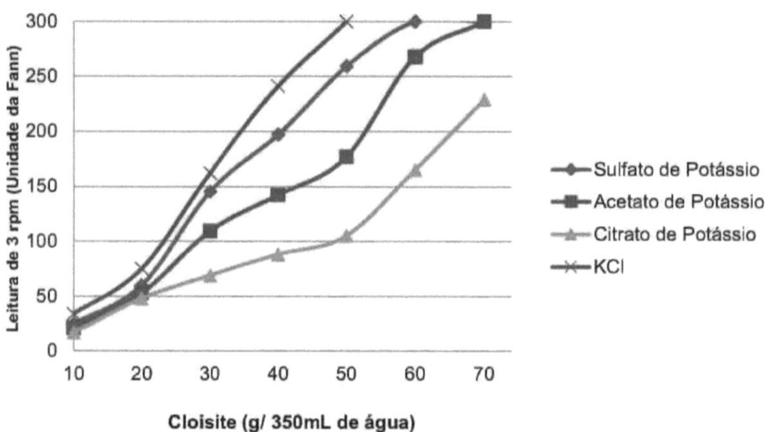

Table 11 - Viscometer readings at 3 rpm for the bentonite inhibition test in relation to Cloisite clay for the inhibitors potassium sulphate, potassium acetate, potassium citrate and potassium chloride.

Concentration	Potassium sulphate	Potassium acetate	Potassium citrate	Potassium chloride
10	25,0	21,0	17,0	34,0

20	60,0	53,0	48,0	75,0
30	145,0	109,0	69,0	162,0
40	197,0	142,0	88,0	241,0
50	259,0	177,0	105,0	300,0
60	300,0	268,0	165,0	300,0
70	300,0	300,0	229,0	300,0

A similar analysis in relation to Figure 55 and Table 11 shows that Cloisite clay behaves similarly to Brasgel PA when compared to hydration inhibitors. However, it is important to emphasize once again that less hydration control by the inhibitors was observed for Cloisite clay compared to that obtained for Brasgel PA clay. This behavior can be explained by the greater reactivity of the former. The solution containing the inhibitor potassium citrate, once again, proved to be more effective in controlling water-clay interaction, which can be seen from the readings for 3 rpm at the clay concentration of 50 g/350 mL of water, which were 259, 177, 105 and 300 rpm, respectively for potassium sulphate, potassium acetate, potassium citrate and potassium chloride. These results also indicate that potassium chloride was the least efficient inhibitor when compared to the other inhibitors.

The ability of an inhibitor to control the incorporation of clay from the formation into the fluid is known to be important, preventing damage to the formation and promoting the maintenance of the fluid's rheological properties. In this way, evaluating the bentonite inhibition capacity is a simple and useful test for evaluating swelling inhibitors in reactive formations. Based on the above, potassium citrate can be identified as the inhibitor that showed the most effective behavior, making it the best inhibitor for controlling expansion in this test.

Based on a joint analysis of the tests carried out to select the best concentrations of clay swelling inhibitors (swelling test, determination of free water by capillary suction and bentonite swelling), it can be suggested that the best concentration to be studied and applied to the drilling fluids to be developed is 20g of inhibitor/350mL of water. It should also be noted that there is a strong indication that potassium citrate will behave as the most effective inhibitor in controlling expansion for the other tests evaluating the inhibitive properties of drilling fluids. This is most likely related to the lower activity of solutions containing this inhibitor compared to the other inhibitors used.

4.3 Determination of the chemical, physical, rheological and filtration properties of drilling fluids

4.3.1 pH measurements of drilling fluids

The pH measurements taken for the drilling fluids inhibited with potassium sulphate, potassium acetate, potassium citrate, potassium chloride and without inhibitor are shown in Figure 56.

Figura 56-   Physical pH values for drilling fluids prepared with the inhibitors potassium sulphate, potassium acetate, potassium citrate and potassium chloride and for the drilling fluid without the inhibitor.

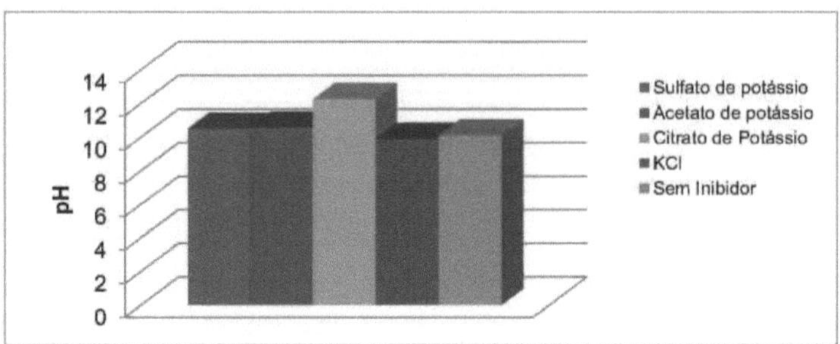

The different pH values obtained for the fluids prepared with the potassium-based inhibitors can probably be explained by the fact that the salts added as expansion inhibitors may or may not be hydrolyzed, since this phenomenon generates a slight increase in pH. Potassium sulphate does not generate hydrolysis, while potassium citrate generates hydrolysis (Magalhaes, 2011). As mentioned above, Figure 56 shows the pH values for the fluids prepared with potassium sulphate, potassium acetate, potassium citrate, potassium chloride and without inhibitor. Based on the pH measurements taken, it can be said that the occurrence of hydrolysis causes the pH of the fluids to change, which can lead to a change in the rheological properties of each of the fluids, thus explaining the possibility of changes in viscosity depending on the nature of the inhibitors used.

According to Lucena et al. (2011c), pH is an influential factor in properties related to fluid performance from values above 10. Thus, fluids with pH values higher than this are likely to show changes in rheological performance.

4.3.2 Density measurements

The density measurements taken for the drilling fluids inhibited with potassium sulphate, potassium acetate, potassium citrate, potassium chloride and without inhibitor are shown in Figure 57.

According to Carvalho and Santos (2004), the drilling fluid is designed to avoid accidents, aiming to counterbalance the natural pressure of the rock formations. As already mentioned, an appropriate balance must be achieved, in which the pressure of the drilling fluid against the walls of the well is sufficient to counterbalance the pressure exerted by the rock formations, but this value must not be too high so as not to damage the well. The pressure of fluids depends basically on their density. Therefore, the results presented by each fluid in relation to this property directly influence its performance.

Figura 57- Values of the physical property of density of the drilling fluids prepared with the inhibitors potassium sulphate, potassium acetate, potassium citrate and potassium chloride and for the drilling fluid without the presence of the inhibitor.

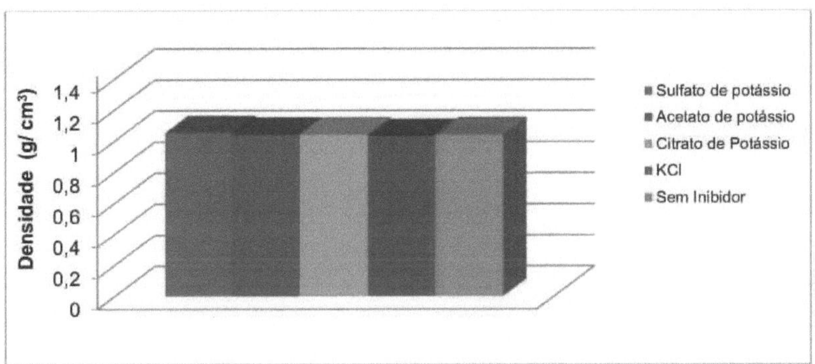

The weight of the fluid and how much pressure it can withstand are known from the density measurement. It is the weight of the fluid, i.e. the pressure it exerts on the formation, that will prevent undesirable fluids from migrating into the well, which is an extremely important factor for the safety of the formation.

It can be seen that the fluids studied have very close density values of 1.04; 1.04; 1.03 and 1.04g/cm^3 , for potassium sulphate, potassium acetate, potassium citrate and potassium chloride, respectively. This conclusion is important because it indicates that the nature of the inhibitor does not have a more incisive influence on the weight of the drilling fluid.

According to a study carried out by Kirschner *et al.* (2011), the density of a polymeric drilling fluid containing the same concentration of calcite used in the formulation of the fluid used in this study showed values between 1.03 and 1.055 g/cm^3 , so the density values obtained were in line with those observed by other authors for fluids with similar compositions. It can therefore be said that the addition of inhibitors does not have a marked influence on this property of the drilling fluid, so it can be used safely in oil well drilling operations.

4.3.3 Rheological and filtration properties

The rheological properties (apparent viscosity (VA), plastic viscosity (VP) and flow limit (LE)) and filtration properties (filtrate volume (VF)) obtained for the polymeric drilling fluids are shown in Figure 58.

Figure 58 shows that the fluids containing the inhibitor have similar rheological properties to the fluids without this additive. This observation is important because, according to Farias *et al.* (2009), the function of the inhibitor is to promote a reduction in the degree of expansion of a reactive formation to be drilled, without significantly altering the properties related to fluid performance. In this way, it can be seen that the presence of the inhibitor generally does not lead to considerable changes in the rheological and filtration properties.

Figura 58-   Rheological and filtration properties of drilling fluids prepared with the inhibitors potassium sulfate, potassium acetate, potassium citrate and potassium chloride and for the drilling fluid without the presence of the inhibitor.

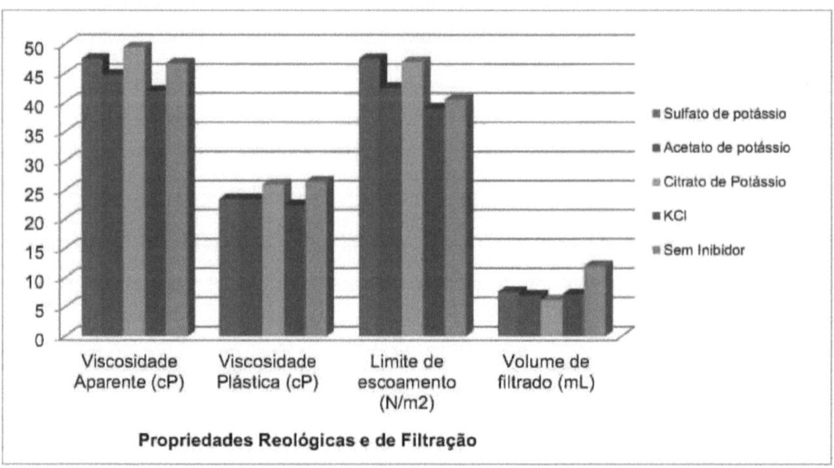

It can be seen that the properties presented by most of the formulated fluids were consistent with those widely used in drilling reactive formations using aqueous fluids, i.e. that the fluids prepared with the chlorine-free expansion inhibitors presented rheological and filtration properties similar to, and in some cases superior to, the fluids prepared with the KCl inhibitor. This behavior may indicate that it is possible to develop an aqueous drilling fluid inhibited with chlorine-free inhibitors with performance comparable to that of commercially used fluids, with fluids prepared with potassium citrate having the best rheological and filtration properties.

In relation to the VF values, it was observed that the chlorine-free inhibited fluids showed results very close to or lower than those presented by the fluid with the same concentration of the inhibitor KCl, which indicates that these fluids perform very well in relation to this property.

For the volume of filtrate, it was observed that the fluid inhibited with potassium citrate obtained a low value for the volume of filtrate, which can be explained by the nature of the inhibitor (which has a higher pH value than the pH presented by the fluids made up of the other inhibitors), this higher pH value produces a greater retention of water in the fluid and this results in lower values for the volume of filtrate.

A subtle difference was observed between the rheological and filtration behavior of the chlorine-free fluids, which can be explained by the nature of each of the additives (potassium sulfate, potassium acetate and potassium citrate) that were added as expansion inhibitors, The small difference in behavior may probably be related to the synergistic relationship that occurs between the additives that make up the drilling fluid formulations and each of the inhibitors (LUCENA, 2012b).

4.4 Determining the inhibitive properties of drilling fluids

4.4.1 Dispersibility test

The shale samples (F1 to F13) were subjected to dispersibility tests with fluids containing the inhibitors potassium sulphate, potassium acetate, potassium citrate, potassium chloride and a fluid without the presence of a swelling inhibitor at the concentration established by the tests that determined the optimum concentration of the inhibitors in section 4.2 (20g of inhibitor / 350 mL of water), as can be seen in Figures 59 to 61.

The dispersibility of shales in a drilling fluid is a function of the particle size of the shale, the Viscosity and inhibitive properties of the drilling fluid, increasing with decreasing particle size of the shale and decreasing with increasing Viscosity of the drilling fluid. However, apparent viscosities above 25cP maintain practically constant dispersion rates. The higher the dispersibility value, the greater the water-rock interaction. Therefore, low values of this property are desirable.

Figura 59- Drilling dispersibility test prepared with the inhibitors potassium sulphate, potassium acetate, potassium citrate and potassium chloride and for the drilling fluid without the presence of the inhibitor for the F1, F2, F3 and F4 shales.

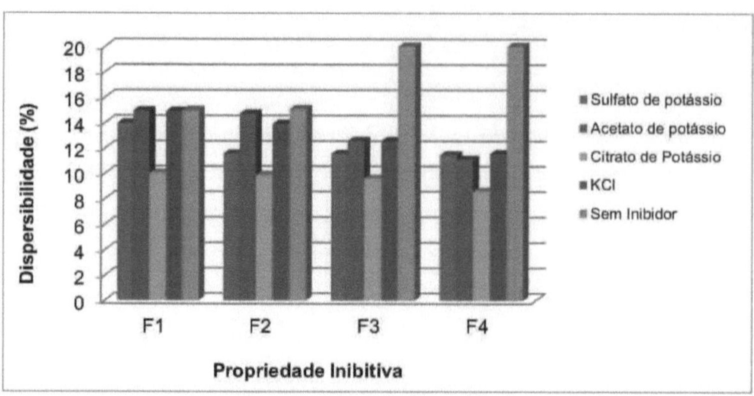

Figura 60- Drilling dispersibility test prepared with the inhibitors potassium sulphate, potassium acetate, potassium citrate and potassium chloride and for the drilling fluid without the presence of the inhibitor for the F5, F6, F7 and F8 shales.

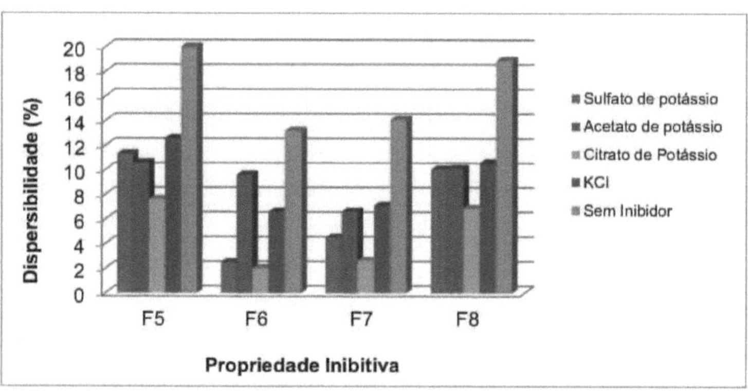

Figura 61- Drilling dispersibility test prepared with the inhibitors potassium sulphate, potassium acetate, potassium citrate and potassium chloride and for the drilling fluid without the presence of the inhibitor for shales F9, F10, F11, F12 and F13.

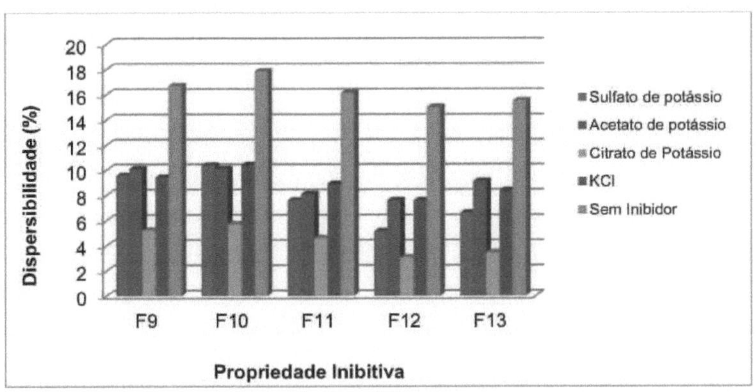

The measured dispersion percentage of fluids contaminated with o shale can be seen in Figures 59 to 61, which show low degrees of shale dispersibility in the presence of fluids containing chlorine-free inhibitors, especially when compared to the values obtained for fluids without the presence of the inhibitor, thus indicating the good performance of these additives in their primary function, which is to control hydration and consequently contain dispersibility.

It was observed that the use of drilling fluid with potassium citrate as an inhibitor in shales resulted in a lower degree of dispersion for all the samples studied, reaching a value of 2.01% for the F6 shale. This result of low dispersibility for this particular shale is consistent with the tests carried out in the course of the work, since the lowest degree of dispersion was obtained for the shale with the lowest reactivity potential and using the fluid with the best performance in expansion control, since according to Suguio (2003) dark shales have a high carbonate content in their composition (as also proven by the characterization carried out), thus presenting a low tendency to swell, i.e. a low dispersibility value, the same behavior is also observed for shale F11.

It can be seen that the activity of shales and fluids is extremely important in determining the flow of water. From this information, it can be seen that the presence of the inhibitor leads to an increase in the concentration of salt in the fluid, which causes it to have a lower water activity than that observed in the formation, thus promoting the flow of the formation into the fluid, which results in a decrease in dispersibility with the use of the swelling inhibitor.

By correlating the results obtained for the tests that evaluate the inhibitive properties with the characterization of the samples studied, it can be seen that the samples with the highest CTC and AE values and which showed a smectite-type clay content (by means of the ATD,

TG, FRX and DRX tests) in their composition were those with the highest dispersibility values (samples F1, F2, F3, F4, F5 and the Brasgel PA and Cloisite clays). This result is in line with the studies and analyses carried out, since shales are geological formations made up of high levels of clays which tend to hydrate, and it is therefore expected that they will swell more clearly for samples in which the water-sensitive material content is higher.

From the above, it can be seen that samples F1, F2, F3, F4 and F5 (Figures 59 and 60) had the highest dispersibility values. The test correlated with the characterization carried out for each of these samples, i.e. the samples with the highest levels of hydratable clay minerals were the ones with the slightly higher dispersibility.

The dispersibility values presented by those samples indicated above as the most reactive can be analyzed using Table 12.

Table 12 - Dispersibility (in %) for samples F1, F2, F3, F4 and F5 in the presence of drilling fluids with the inhibitors potassium sulphate, potassium acetate, potassium citrate and potassium chloride.

SAMPLE NAME	Potassium sulphate	Potassium acetate	Potassium citrate	Potassium chloride	No inhibitor
F1	13,99	14,44	10,04	14,91	23,99
F2	11,59	14,72	9,87	13,91	23,07
F3	11,56	12,59	9,60	12,59	21,59
F4	11,48	11,10	8,61	11,59	20,34
F5	11,34	10,59	7,60	12,58	19,49

The data in Table 12 shows that potassium citrate once again proved to be the most effective in controlling the swelling of the samples in general. This result confirms that the effectiveness of the potassium citrate inhibitor may be linked to its various characteristics, such as its higher pH value, which increases the rheology of the fluid and consequently reduces the interaction between the shale and the fluid, and that the high hydration control capacity of this inhibitor may also be due to the physical inhibition promoted by its counter-ion, which has three anionic active sites.

It can also be seen that the dispersibility of samples F6, F7, F8, F9, F10, F11, F12 and F13 is also consistent with the characterization of the samples, since they have lower dispersibility values than samples with reactive clay mineral content in their composition, However, it is important to note that the use of inhibitor fluids led to a reduction in the dispersibility values of these samples, which may be an indication that the inhibitors are not only efficient in controlling hydration, but are also effective in reducing the dispersibility of non-reactive samples.

Chapter 5

5. CONCLUSIONS

With the aim of studying and evaluating the efficiency of inhibited and chlorine-free aqueous drilling fluids in controlling the hydration of samples of shale formations from various regions of the country, it was concluded that:

•   according to the characterization of the samples studied, the shales that indicate the presence of smectite in their composition, according to the XRF and XRD tests, showed moderate to high reactivity according to the cation exchange capacity test, specific area, as well as thermal and thermogravimetric behavior similar to that shown by the clay minerals of the smectite group;

•   the characterization tests on the shales showed that the F1, F2, F3, F4 and F5 shales, and the reactive clays Brasgel PA and Cloisite all have characteristics that indicate the reactivity of the samples;

•   the tests evaluating the efficiency of the chemical inhibitors showed that the concentration with the best results for all the inhibitors studied was 20g/350mL of water and that the potassium citrate inhibitor showed the best performance. it can also be said that chlorine-free inhibitors showed superior results compared to the performance of the inhibitor containing chlorine in its composition;

•   fluids prepared with chlorine-free expansion inhibitors showed similar rheological and filtration properties and in some cases were superior to fluids prepared with the KCl inhibitor;

•   The fluid prepared with the inhibitor potassium citrate showed the lowest degree of dispersion for all the samples studied. This performance can be attributed to the various characteristics it has, such as a higher pH value, which increases the fluid's rheology, reducing the interaction between the shale and the fluid, as well as the high hydration control capacity of this inhibitor, attributed to the physical inhibition promoted by its counter-ion, which has three anionic active sites,

•   the characterization of the samples made it possible to determine the reactive potential of the samples studied and acted as an indispensable tool in assessing the behaviour of the samples in the face of hydration, correlating perfectly with the dispersibility test carried out, thus concluding, that the characterization of the samples acts as an indicator of their susceptibility to swelling, and this indicator is confirmed by the relationship established with the inhibitive property, since the samples that showed characterization with indications of the presence of reactive clay minerals were the ones that showed the greatest sensitivity to the presence of water.

SUGGESTIONS FOR FUTURE WORK

With a view to contributing to future research that will allow the work to continue, the following points are suggested:

- Further studies into the swelling of reactive formations by developing a system that simulates the conditions of shales in a well, both in terms of temperature and pressure. This is necessary in order to quantify the effects of these variables on shale-fluid interaction.

- Expand studies with other chlorine-free inhibitor additives in order to increase the range of products intended for this function;

- Carry out the bentonite swelling test using drilling fluids in contact with the samples;

- Evaluation of inhibited fluids through the study of the degree of swelling of shales by *Linear Swell Meter* tests for the compositions of inhibited fluids developed;

- Evaluate the dispersibility control action of the inhibitors studied and,

- To verify the biodegradability of the inhibitors studied in order to prove the environmental safety of chlorine-free inhibited fluids.

BIBLIOGRAPHICAL REFERENCES

ALBANO, G. C., FAGUNDES, F. P., DA SILVA, Í. G. M., ARAÙJO, B. A. B. D., BALABAN, R. C., **Evaluation of the kinetic swelling parameters of bentonite clays in aqueous fluids**, Proceedings of the 6th⁰ PDPETRO, Florianopolis, SC, 2011.

AL-BAZALI, T. M., **The consequences of using concentrated salt solutions for mitigating wellbore instability in shales**, Journal of Petroleum Science and Engineering, v. 80, p. 94-101, 2012

AL-BAZALI, T. M., AL-MUDHI, S., CHENEVERT, M. E., **An Experimental Investigation of the Impact of Diffusion Osmosis and Chemical Osmosis on the Stability of Shales**, Journal of Petroleum Science and Technology, v. 29, p. 312-323, 2011.

ALBUQUERQUE, A. C. C., VIANNA, A. M.; PENNA, M. O., SOUZA, L. S., OLIVEIRA, R. M. S., KRUGER, V. A. N., BRANCO, M. A., VALERIO, R. R., MOTA, V. C., **Biodegradation of Drilling Fluid X Operating Conditions,** 26⁰ Congresso Brasileiro de Microbiologia, Foz do Iguaçu, PR, 2011.

AMORIM, C. L. G.; LOPES, R. T.; BARROSO, R. C.; QUEIROZ, J. C.; ALVES, D. B.; PEREZ, C. A.; SCHELIN, H. R., **Effect of clay-water interactions on clay swelling by X-ray diffraction**. Nuclear Instruments & Methods in Physics Research , v. 580, p. 768-770, 2007.

AMORIM, L.V., CAMPOS, L. F. A., FERREIRA, H. C., **Using experimental planning to study the effect of the composition of bentonite mixtures on the rheology of drilling fluids. Part I: binary compositions**. Ceràmica (Sâo Paulo. Impresso), Sâo Paulo, v. 52, n.321, p. 69-75, 2006.

AMORIM, L. V., GOMES, C. M., SILVA, F. L. H., LIRA, H. L., FERREIRA, H. C., **Rheological study of water-based drilling fluids: Influence of solids content, speed and agitation time**, Aguas Subterràneas, v. 19, n. 1, p. 75-85, 2005.

ANDERSON, R.L., RATCLIFFE, I., GREENWELL, H.C., WILLIAMS, P.A, CLIFFE, S.P.; COVENEY, V., **Clay swelling - A challenge in the oilfield**, Earth-Science Reviews, n⁰ 98, p. 201-216, 2010.

API, **Petroleum and natural gas industries - Field testing of drilling fluids. Part 1: water-based fluids**, 2005.

API, **API Standard Recommended Practice 13B-1**, November 2003.

BAILEY, L., KEALL, M., AUDIBERT, A., LECOURTIER, J., **Effect of clay / polymer interactions on shale stabilization during drilling,** Langmuir, v. 10, p. 1544-1549, 2004.

BALTAR, C. A. M., DA LUZ, A. B., **Mineral inputs for drilling oil wells,** ISBN: 85-7227-187-2 - Rio de Janeiro: CETEM / UFPE, 91 p, 2003.

BASSI, G.L., FEDERICI, F.T.B.; BOSSI, T., MERLI, L., VIGANO, L., BOTTARELLO, L., **Swelling Inhibitors for Clays and Shales**, Patent application USPC Class: 507131, Houston, U.S.A, 2009.

BATISTA, A. P. S.; MENEZES, R. R.; MARQUES, L. N.; CAMPOS, L. A.; NEVES, G. A.; FERREIRA, H. C. **Caracterizaçao** de **argilas bentoniticas** de **Cubati - PB**, Revista Eletronica de Materials e Processos, v.4.3 (2009) 64-71.

BLACHIER, C., MICHOT, L., BIHANNIC, I., BARRES, O., JACQUET, A., MOSQUET, M., **Adsorption of polyamine on clay minerals**, Journal of Colloid and Interface Science, n^0 336, p. 599-606, 2009.

BOURGOYNE Jr, A.T., MILLHEIM, K.K., CHENEVERT, M.E., YOUNG Jr, F.S., **Applied drilling engineering**. 2 Ed. Richardson, Texas, Society of Petroleum Engineers,1991.

CAMPOS, L. F. A. **Bentonite clay compositions for use in oil well drilling fluids**, Doctoral thesis presented to the Process Engineering Course/CCT/UFCG, 2007.

CARDOSO, J. J. B., **Study of the swelling of sodium bentonites and evaluation of the performance of inhibitors by X-ray diffraction**, 127 f. Dissertation (Master of Science in Nuclear Engineering) - Federal University of Rio de Janeiro, Rio de Janeiro/RJ, 2005.

CARTER D. J., OGDEN, M. I., ROHL, A. L., **Mechanisms of dye incorporation into potassium sulfate: Computational and experimental studies**, Journal of Physical Chemistry C, v. 111, p. 9283-9289, 2007.

CARVALHO, A. L., SANTOS, J. N., **Processing of drilling muds water-based muds and oil-based muds**, IV Workshop do PRH 16, Itajuba, MG, 2004.

CELIK₁ H., **Technological characterization and industrial application of two Turkish clays for the ceramic industryı** Applied Clay Scienceı v. 50ı p. 245-254ı 2010.

CHENG, H., LIU, Q., ZHANG, J., YANG, J., FROSTi R. L., **Delamination of kaolinite-potassium acetate intercalates by ball-milling**, Journal of Colloid and Interface Science, v.348, p. 355-369, 2010.

CHESSER, B.G., **Design considerations for an inhibitive, stable water-based mud system.**SPE Paper 14757, IADC/SPE Drilling Conference, 1987.

CHURCHMAN, G.J., **Formation of complexes between bentonite and different cationic polyelectrolytes and their use as sorbents for non-ionic and anionic pollutants**, Applied Clay Science, vol. 21, 2002.

CUNHA, F. W., PRETTO, T. R., MENEZES, E. W., CALDAS, E. M., BENVENUTTI, E.V., COSTA, T. M. H., DIAS, S. L. P.; ARENAS, L. T., **Synthesis of ceramic carbon containing silver nanoparticles and its application in the reduction of hydrogen peroxide**, I Workshop on functional hybrid materials, Rio Grande do Sul, RS, 2013.

CYGAN, R.T., GREATHOUSE, J.A., HEINZ, H., KALINICHEV, A.G., **Molecular models and simulations of layered materials**. Journal of Materials Chemistry, v.9, p. 2470-2981, 2009.

DARLEY, H.C.H., GRAY, G.R., **Composition and Properties of Drilling and Completion Fluids.** Gulf Publishing Company, SPE/ADC Paper 92367, SPE/ADC Drilling Conference.Fifth Edition, 1988

DIAZ-PEREZ, A., CORTES-MONROY, I., ROEGIERS, J.C., **The role of water/clay interaction in the shale characterization**, Journal of Petroleum Science and Engineering, n^0 58, p. 83-98., 2007

DOMINIJANNI, A., MANASSERO, M., **Modelling the swelling and osmotic properties of clay soils. Part I: The phenomenological approach**. International Journal of Engineering Science, v. 51, p. 32-50, 2012.

DURAND, C., FORSANS, T., RUFFET, C., AUDIBERT, A., Influence of clays on borehole stability, **Reveu de L'Institute Français du Pétrole**, Vol. 50, no 2, 1995.

DYE, W., D'AUGEREAU, K., HANSEN, N., OTTO, M., SHOULTS, L., LEAPER, R., CLAPPER, D., XIANG, T., **New water-based mud balances high-performance drilling and environmental Compliance**, p. 23-25, 2006.

ERZIN, Y., EROL, O., **Swell pressure prediction by suction methods**, Engineering Geology, v. 92, p. 133-14, 2007.

EWY, R. T.; STANKOVICH, R. J., **Shale-Fluid Interactions Measured Under Simulated**

Downhole Conditions. SPE/ISRM Rock Mechanics Conference, SPE 78160, Irving, 10 p, 2002.

FARIAS, K.V., AMORIM, L. V., LIRA, H. L., **Development of aqueous fluids for use in oil well drilling - Part I,** Electronic Journal of Materials and Processes (UFCG), v. 4.1, p. 14-25, 2009.

FERREIRA, H.S., MENEZES, H. S., FERREIRA, H. S., MARTINS, A. B., NEVES, G. A., FERREIRA, H. C., **Analysis of the influence of purification treatment on the swelling behavior of organophilic clays in non-aqueous media,** Ceràmica, v. 54, p. 77-85, 2008.

FERREIRA, H.C., CHEN, T., ZANDONADI, A.R. & SOUZA SANTOS, P., **Linear Correlations between Specific Areas of Kaolins Determined by Various Methods - Application to Some Kaolins from the Brazilian Northeast (States of Paraiba and Rio Grande do Norte),** Ceràmica 18 (71) 333 (1972).

FERREIRA, H. S. **Optimization of the Organophilization Process of Bentonites for Use in Non-Aqueous Drilling Fluids,** Doctoral Thesis presented to the Materials Engineering Department of the Federal University of Campina Grande, November 2009.

FONTOURA, S. A. B., **Lade and Modified Lade 3D Rock Strength Criteria. Rock Mechanics and Rock Engineering,** v. 45, p. 1001-1006, 2012.

FONTOURA, S. A. B., **Geotechnical Behavior of Sedimentary Argillaceous Rocks**. 5 th ISRM International Symposium. Proceedings of the 5 th Asian Rock Mechanics Symposium, v. 1. p. 59-72, 2009.

GANG LU, J., CHUN-HUA, A., WEN XU, Z., FAN, F., CHENG, L., YING LIN, F., **Measurement and prediction of densities, viscosities, and surface tensions for aqueous solutions of potassium citrate,** Fluid Phase Equilibria, v. 327, p. 9- 13, 2012.

GRAY, G. R., DARLEY, H.C.H., **Composition and properties of oil well drilling fluids,** 1981.

GHASSEMI, A., DIEK1 A., SANTOS, H., **Influence of coupled chemo-poro-thermoelastic processes on pore pressure and stress distributions around a wellbore in swelling shale,** Journal of Petroleum Science and Engineering, v.67, p. 57-64, 2009.

GHOLIZADEH-DOONECHALY, N., TAHMASBI, K., DAVANI, E., **Development of high performance water-based mud formulation based on amine derivatives,** SPE Paper 121228, SPE International Symposium on Oilfield Chemistry. Woodlands, p. 20- 22, 2009.

HAWKIES, C. D., McLELLAN, P. J., RUAN, C; MAURER, W., **Wellbore Instability in**

Shales: A Review of Fundamental Principles and GRI-Funded Research, GRI Project Manager, 2000.

HORTON, D., VAN, R. V., TANCHE-LARSEN, Surfactant additives used to retain producibility while drilling, CA 2803542 A1, 2012.

HOWARD, S.K., Formate brines for drilling and completion: state-of-the-art. Paper SPE 30498 - SPE Annual Conference and Exhibition, 1995.

ISMAIL, I., HUANG, A. P., The application of methyl glucoside as shale inhibitor in sodium chloride mud, Jurnal Teknologi, 50, p. 53-65, 2009.

JONES, A. G., MULLIN, J. W., Programmed cooling crystallization of potassium sulphate solutions, Chemical Engineering Science,v. 29, p. 105- 118, 1974.

KARABORNI, S., SMIT, B., HEIDUG, W.K., URAI, J.L., VAN OORT, E., The swelling of clays: molecular simulations of the hydration of montmorillonite, Science, v. 271, p. 1102-1104, 1996.

KATTI, K.S., KATTI, D.R., Silica-water interactions in montmorillonite using Fourier transform infrared spectroscopy: relationship to swelling and swelling pressure, Langmuir, n⁰ 22, p. 532-537, 2006.

KEHEW, E.A., Geology for Engineers and environmental scientists, 3ᵃ Edition, Prentice Hall Englewood Cliffs, New Jersey, 2006.

KEIJZER, .J.S., HEISTER, K, LOCH, J.P.G., Stability of clay membranes in chemical osmotic processes, Clays and Clay Minerals, v. 169, p. 632-639, 2004.

KELLER, E.A., Environmental Geology, Eighth edition. Prentice Hall, USA, 2008.

KHODJA ,M., CANSELIER, J. P., BERGAYA, F., FOURAR, K.; KHODJA, M., COHAUT, N., BENMOUNAH, A., Shale problems and water-based drilling fluid optimization in the Hassi Messaoud Algerian oil field, Applied Clay Science, v. 49, p 383-393, 2010.

KOMINE, H., Predicting hydraulic conductivity of sand-bentonite mixture backfill before and after swelling deformation for underground disposal of radioactive wastes, Engineering Geology, no 114, p. 123-134, 2010.

KIRSCHNER, B. D., BARROS NETO, E. L., DANTAS NETO, A. A., Evaluation of the Stability of Water-Based Drilling Fluids, 6th Brazilian Congress on R&D in Oil and Gas, 2011, Florianopolis, 2011.

LAST, N., PLUMB, D., Managing Wellbore Stability in the Cusiana Field. The Search, for Oil & Gas in Latin America & the Caribbean, Slumberger Surenco CA., No. 2, pp. 8-31, 1995.

Report: High performance fluid project, requested by the company System Mud Ind. Comercio LTDA, carried out by LAPET/ UFRN, 2011.

LEAL, C. A., NASCIMENTO, R. C. A. M., AMORIM, L. V., **Study of bentonite suspensions under different thermal conditions**. Ceràmica Industrial, v. 59, p. 115-123, 2013.

LEITE, I.F., RAPOSO, C. M. O., SILVA, S. M. L., **Structural characterization of national and imported bentonite clays: before and after the organophilization process for use as nanocharges**, Revista Ceràmica, v. 54, p. 303-308, 2008.

LOMBA, R. F. T., SA, H. M. S., PEREZ, R. C., **Development of a test methodology to evaluate the interaction between shale and drilling fluid,** Organic Geochemistry, PETROBRAS, 2000.

LONG, Y., LIN, Z., XIA, M., ZHENG, W., LI, .Z., **Mechanism of HERG potassium channel inhibition by tetra-n-octylammonium bromide and benzethonium chloride**, Toxicology and Applied Pharmacology, v. 267, p. 155-166, 2013.

LOPES, L. F., SILVEIRA, B. M. O., MORENO, R. B. Z. L., **Loss circulation and formation damage control on overbalanced drilling applying different formulations of water based drill-in fluids on sandstone reservoir**, 31st International Conference on Ocean, Offshore and Arctic Engineering OMAE, Rio de Janeiro, RJ, 2012.

LUCENA, D.V., AMORIM, L. V.; LIRA, H.L., **Analysys of the influence of lubricant and sealant in inhibited fluids,** 22nd International Congress of Mechanical Engineering- COBEM 2012, Ribeirao Preto, SP, 2012a.

LUCENA, D. V., AMORIM, L. V., LIRA, H.L., **Influence of the inhibitor on the performance of environmentally friendly polymeric drilling fluids.** Brazilian Congress of Materials Science and Engineering - CBECIMAT 2012, Joinville, 2012.

LUCENA, D.V., AMORIM, L. V.; LIRA, H.L., **Development of chlorine-free inhibited drilling fluids,** IV Encontro Nacional de Hidraulica de Poços de Petroleo e Gas- ENAHPE 2011, Foz do Iguaçu, PR, 2011a.

LUCENA, D.V., SANTANA, F. R., AMORIM, L. V., BARBOSA, J.A., **Development of inhibited fluids for drilling shales in the Araripe Basin**, 6^0 Congresso Brasileiro de Petroleo e Gas, 6^0 PDPETRO, Florianopolis, SC, 2011b.

LUCENA, D.V., AMORIM, L. V.; LIRA, H.L., **Development of chlorine-free inhibited**

drilling fluids, Revista Petro & Quimica, v. 334, 2011c.

LUMMUS, J.L., AZAR, J.J., **Drillings fluids optimization a practical field approach**, Penn Well Publishing Company, Tulsa, Oklahoma, 1986.

MACHADO J. C. V., **Rheology and fluid flow**. Editora Interciència, Rio de Janeiro, 2002.

MAGALHÂES, J. , **Drilling fluids: economy and high performance. System Mud. Drilling fluids**. Available at: http://www.svstemmud.com.br/artiqos/fluidosde perfuracao %20economia e alto desempenho.pdf, accessed on: 01/11/2013.

MARTINS, A. B., FERREIRA, H. S., FERREIRA, H. C., NEVES, G. A., **Development of organized bentonite clays for use in non-aqueous fluids with low aromatic content,** Anais do 4° PDPETRO, Campinas, SP, October, 2007.

MANOHAR, L., **Shale Stability: Drilling Fluid Interaction and Shale Strength**, SPE 54356, SPE Latin American and Caribbean Petroleum Enqineerinq Conference, p. 21-23,1999.

MELO JR., M. A., SANTOS, L. S. S., GONÇALVES, M. C., NOGUEIRA, A. F., **Preparation of silver and gold nanoparticles: a simple method for introducing nanoscience in teaching laboratories**, Quimica Nova, v. 35, n° 9, p., 2012.

MELÉNDEZ, V. M. A., **Avaliação Experimental dos Parâmetros de Transporte em Folhelhos**, Civil Engineering Dissertation, Pontificia Universidade Catolica - PUC- Rio, 184p, 2010.

MENEZES, R. R., CAMPOS, L. F. A., FERREIRA, H. S., MARQUES, L. N., NEVES, G. A., FERREIRA, H. C, **Estudo do comportamento rheológico das argilas bentoniticas de Cubati, Paraiba, Brasil,** Ceràmica, v. 55, p. 349-355, 2009.

MONTILVA, J., OORT, E., BRAHIM, R., LUZARDO, J.P., MCDONALD, M., QUINTERO, L., DYE, B., TRENERY, J., **SPE-110366**, SPE Annual Technical Conference and Exhibition, Anaheim, U.S.A. 2007.

MOKNI, N., OLIVELLA, S., ALONSO E.E., **Swelling in clayey soils induced by the presence of salt crystals**, Applied ClavScience, v. 47, p. 105-112, 2010.

MOTTA, J. F. M., LUZ₁ A.B., BALTAR, C. A. M., BEZERRA, M. S., CABRAL JÛNIOR, M., COELHO, J. M., **Plastic clay for white ceramics**, Industrial Rocks and Minerals, v. 9, p. 33-46, 2008.

MUNIZ, E.S., FONTOURA, SAB, LOMBA, R. F. T., DUARTE R.G., **Evaluation of the**

shale-drilling fluid interaction for studies of wellbore stability. Symposium on Rock Mechanics, 2005.

NASCIMENTO, R. C. A., AMORIM, L. V., LIRA, D. S., LIRA, H. L., **O Fenômeno de Prisâo Diferencial: Uma revisâo da Literatura**, Revista Eletronica de Materiais e Processos (UFCG), v. 5.2, p. 76-87, 2010a.

NASCIMENTO, R. C. A. M., AMORIM, L. V., SANTANA, L. N. L., **Development of aqueous fluids with bentonite for drilling *onshore* oil wells**, Ceràmica, v.56, p.179-187, 2010b.

NASCIMENTO, R. C. A., VIEIRA, T.M., AMORIM, L. V., LIRA, H.L., **Evaluation of the efficiency of expansive clay inhibitors for use in drilling fluids**, Electronic Journal of Materials and Processes (UFCG), v. 4.2, p. 12-19, 2009.

NIU, M., WANG, S., HAN, X., JIANG, X., **Yield and characteristics of shale oil from the retorting of oil shale and fine oil- shale ash mixtures**, Applied Energy, v. 111, p. 234-239, 2013.

NORRISH, K., RAUSELL-COLOM, J. A., **Effect of freezing on the swelling of clays minerals**, Clay Miner. V. 5, p. 9-16, 1963.

O'BRIEN, D. E., CHENEVERT, M. E., **Stabilizing sensitive shales with inhibited, potassium-based drilling fluids**, Journal Petroleum Technology, p. 1089-1100, 1973.

OSISSANYA, S.O., ENILARI, M.G., AYENI, K.B., **Development and Evaluation of Various Drilling Fluids for Slim Hole Wells**, Journal of Canadian Petroleum Technology, v. 48, p. 30-32, 2009.

PATEL, A., STAMATAKIS, E., YOUNG, S., FRIEDHEIM, J., **Advances in inhibitive water based drilling fluids- Can they replace Oil based muds?**, SPE/ Internation Syposium on Oilfield Chemistry, Texas, N° 28960, 1995.

PEREIRA, E. *et al*, **Uso de Inibidores de Argilas como Soluçâo de Problemas em Sondagem**, 2007, Available at: www.systemmud.com.br, Accessed on: November, 2012.

PETIT, S., **Fourier Transform Infrared Spectroscopy**, Lagaly, G. (Ed.), Handbook of ClayScience. Developments in ClayScience, vol. 1. Elsevier, p. 909-918, 2006.

POWELL, J.S., SIEMENS, G.A. TAKE, W.A., REMENDA, V.H., **Characterizing the swelling potential of Bearpaw clayshale**, Engineering Geology, v. 158, p. 8997, 2013.

PRADO, C.M.O., JUNIOR, F.S., NASCIMENTO, J.M., BARRETO, C.A., BERTOLINO, L.C., FRANÇA, S.C.A., PATROCINIO, J., **Characterization of clays used in the production of red ceramics in the State of Sergipe**, 560 Brazilian Congress of Ceramics, 10 Latin

American Congress of Ceramics, p. 745-755, 2012.

PUPPALA, A.J., CERATO, A.B., **Heave distress problems in chemically-treated sulfate laden materials.** GeoStrata, v. 102, p. 28-32, 2009.

QU , Y., LAI, X. Q., ZOU, L., SU, Y., **Polyoxyalkyleneamine as shale inhibitor in water-based drilling fluids**, Applied Clay Science, v. 44, p. 265-268, 2009.

RABE, C., CHERREZ, J. O., **Laboratory Characterization of Norwegian North Sea Shale. In: Laboratory Characterization of Norwegian North Sea Shale, 5th Asian Rock Mechanics Symposium**, 2009. v. 1.p. 37-44, 2009.

RABE, C., FONTOURA, S. A. B., **Effect of organic salts on the physicochemical properties of shales.** Brazilian Oil and Gas Congress - Rio Oil and Gas. Rio de Janeiro, 2003.

RICHARD, L. A., GREENWELL, H.C., SUTER, J.L., JARVIS, R.M., COVENEY, P.V., Annals of the Brazilian Academy of Sciences, n⁰ 82(1), 43, 2010.

RODRIGUES, M. G. F., SILVA, M. L. P., SILVA, M. G. C., **Characterization of bentonite clay for use in the removal of lead from synthetic effluents**, Ceràmica, Vol. 50, p. 190-196, 2004.

ROSA, R. C., FARIAS, A. L., GARCIA, S. B., COELHO, M. H., **A new inhibitive water-based fluid: A completely cationic system**, SPE Latin American and Caribbean Petroleum Engineering Conferece, Rio de Janeiro, Brazil, 2005.

SANTOS, M. I., GARCIA, R. B., GIRÂO, J. H. S., SANTOS, K. P. F., Vidal, E. L. F., COSTA, R. M. C., SANTOS, T. L. P. S., **Performance Evaluation of Emulsion Preventers for Completion and Stimulation Fluids.**

Evaluation of the Performance of Demulsifiers for Completion and Stimulation fluids, Rio Oil and Gas. Rio de Janeiro, RJ, 2012.

SANTOS, H. M., **Method for the evaluation of shale reactivity,** Patent application USPC Class: 6247358 B1, U.S.A, 2001.

SANTOS, T. T., AMORIM, L.V., **Study of the filtration properties of aqueous drilling fluids**, 76 f. Dissertation (Master's Degree in Materials Science and Engineering) - Federal University of Campina Grande/PB, 2013.

SERRA, A. C. S., SANTOS NETO, E. V., **The influence of drilling mud additives on the geochemical properties of oils**, Revista Petroquimica, Petroleo, Gas & Quimica, v. 1,p. 98-102, 2004.

SHUIXIANG, X., GUANCHENG, J., MIAN, C., **An environmentally friendly drilling fluid system.Petroleum Exploration and Development**, v. 38 (3), p. 369-378, 2011.

SILVA, G. V., **"Roteiro de testes para avaliaçâo do caratera de inibiçâo dos sal de chloretos e polimeros utilizados nas operações de perfuraçâo de poços de petróleo e gás"**. Natal, UFRN, November 2005.

SILVA, I. A., COSTA, J.M.R., NEVES, G.A., FERREIRA, H.C., **Organophilization of bentonite clays with non-ionic surfactants for application in drilling fluids**. Ceràmica, v. 58, p.317- 327, 2012.

SILVA, S. P., ESTEVÂO, L. R. M., CSABA, N., NASCIMENTO, R. S. V., **Clays Basal Spacings Effect on Fire Retardant Properties of Polymeric Materials by**

Thermal Analysis. Journal of Thermal Analysis and Calorimetry, v. 106, p. 535-539, 2011.

SILVA, W. G. A. L., ALMEIDA, R. D. F., **Analysis of physico-chemical parameters of drilling fluids with polymer additives**, Rio Oil and Gas, Rio de Janeiro, 2010.

SIMPSON, J.P., DEARING, H.L., **Diffusion Osmosis - An Unrecognized Cause of Shale Instability**, IADC/SPE 59190, Drilling Conference held, New Orleans, 2000.

SOUZA, B. B., BORGES, S. P. M. S., **Influence of sodium and potassium chlorides on plastic properties and mechanical behavior for cementing oil wells**, Graduation Project, Graduation in Petroleum Engineering, UFRJ, Rio de Janeiro, 2011.

SOUZA SANTOS, P., **Ciência e tecnologia de argilas**, vol.2, Editora Edgard Blucher Ltda., Sâo Paulo, 1992.

Standard Test Method for Swell Index of Clay Mineral Component of Geosynthetic Clay Liners - ASTM D 5890-11.

STEFAN, P., **Effect of Na, K, Ca, Mg, Ba and Sr salts on the properties of drilling mud in deep oil and rock salt drilling**. 3. ed. Sergipe/SE: Livraria Regina LTDA, 1956.

SUGUIO, K., Dicionario de geologia sedimentar e áreas afins. BDC Uniao de Editoras S.A., 1998.

SUGUIO, K., **Sedimentary Geology.** Sao Paulo: Editora Edgard Blücher, 2003.

SUTER, J.L., ANDERSON, R.L., GREENWELL, H.C., COVENEY, P.V., **Recent advances in large scale atomistic and coarse-grained molecular dynamics techniques applied to clay minerals**. Journal of Materials Chemistry v.19, p. 24822493, 2009.

SUTER, J.L., COVENEY, P.V., GREENWELL, H.C., THYVEETIL, M.A., **Large- scale** molecular dynamics study of montmorillonite clay: emergence of undulatory fluctuations and determination of materials properties, Journal of Physical Chemistry, n⁰ 111, p. 8246-8259, 2007.

TAN, C. P., RAHMAN, S. S., CHEN, X., **Wellbore Stability Analysis and Guidelines for Efficient Shale Instability Management.** Paper SPE/ADC 47795, SPE/ADC Drilling Conference, Indonesia, 1998.

THOMAS J.E., **Fundamentos de engenharia de petróleo**, Editora Interciência, Rio de Janeiro, 2001.

VAN OORT, E., **On the physical and chemical stability of shales**, Journal of Petroleum Science and Engineering, v. 38, p. 213- 235, 2003.

VIDAL, E.L.F., FELIX, T.F., GARCIA, R.B., COSTA, M., GIRÂO, J.H.S., **Application of new cationic polymers as clay inhibitors in water-based drilling fluids**, Anais do 4⁰ PDPETRO, Campinas, SP, October, 2007.

VIDAL, E.L.F., GARCIA, R.B., GIRÂO, J.H.S., COSTA, M., SANTOS, M. I.., SANTOS, K. P. F., COSTA, R. M., SANTOS, J. A. C. M., FELIZARDO FILHO, A., **Thermal resistance evaluation of vegetable oil-based drilling fluids**, Rio Oil and Gas Expo & Conference, Rio de Janeiro, 2006.

VILLAR, M.V., GÓMEZ-ESPINA, R., **Report on thermo-hydro-mechanical laboratory tests performed by CIEMAT on FEBEX bentonite**, Informes Técnicos CIEMAT 1178, Madrid, 2012.

WARR, L.; BERGER, J., **Hydration of bentonite in natural waters: application of "confined volume" wet-cell X-ray diffractometry**, Physics and Chemistry of the Earth, n⁰ 32, p. 247-258, 2007.

XUAN, Y.; JIANG, G.; LI, Y.; GENG, H.; WANG, J., **A biomimetic drilling fluid for wellbore strengthening Petroleum Exploration and Development**, Petroleum Exploration and Development, v. 40 (4), p. 531-536, 2013.

YAN, C., DENG, J., YU, **Welbore stability in oil and gas with chemicalmechanical coupling,** ScientificWorld Journal v.2013, 2013.

YANG, L., PHUA, S. L., TEO, J. K. H., **A biomimetic approach to enhancing interfacial interaction: Polydopamine- coated clay as reinforcement for epoxies resins**, ACS Apllied Material & Interfaces, v. 3 (8), p.3032, 2011.

YE, W. M., CHEN, Y.G., CHEN, B., WANG, Q.; WANG, J., **Advances on the knowledge** of the buffer/backfill properties of heavily-compacted **GMZ** bentonite, Engineering Geology, v. 116, p. 12-20, 2010.

ZAMPORI, L.,DOTELLI, G.,GALLO STAMPINO, P., ZORZI, F., FINOCCHIO, E., **Thermal characterization of a montmorillonite, modified with polyethyleneglycols (PEG1500 and PEG4000), by in situ HT-XRD and FT IR: Formation of a high-temperature phase**, Applied Clay Science, 2012.

ZHANG, J., AL-BAZALI; T. M. , CHENEVERT, M. E., SHARMA; M., **Factors controlling the membrane efficiency of shales when interacting with waterbased and oil-based muds**, SPE Drill, Completion, v. 23 (2), p. 150-158, 2008.

ZHONG, H., ; QIU ,Z., HUANG, W., CAO, J., , **Shale inhibitive properties of polyether diamine in water-based drilling fluid**, Journal of Petroleum Science and Engineering, v. 78,p. 510-515, 2011.

yes
I want morebooks!

Buy your books fast and straightforward online - at one of world's fastest growing online book stores! Environmentally sound due to Print-on-Demand technologies.

Buy your books online at
www.morebooks.shop

Kaufen Sie Ihre Bücher schnell und unkompliziert online – auf einer der am schnellsten wachsenden Buchhandelsplattformen weltweit! Dank Print-On-Demand umwelt- und ressourcenschonend produziert.

Bücher schneller online kaufen
www.morebooks.shop

info@omniscriptum.com
www.omniscriptum.com

Printed by Books on Demand GmbH, Norderstedt / Germany